William Dell Hartman

Conchologia cestrica

The molluscous animals and their shells of Chester county

William Dell Hartman

Conchologia cestrica
The molluscous animals and their shells of Chester county

ISBN/EAN: 9783337228651

Printed in Europe, USA, Canada, Australia, Japan

Cover: Foto ©berggeist007 / pixelio.de

More available books at **www.hansebooks.com**

CONCHOLOGIA CESTRICA.

THE

MOLLUSCOUS ANIMALS

AND THEIR

SHELLS,

OF CHESTER COUNTY, PA.

BY

WILLIAM D. HARTMAN, M.D.,

AND

EZRA MICHENER, M.D.

𝔚𝔦𝔱𝔥 𝔑𝔲𝔪𝔢𝔯𝔬𝔲𝔰 𝔍𝔩𝔩𝔲𝔰𝔱𝔯𝔞𝔱𝔦𝔬𝔫𝔰.

"An undevout philosopher is mad."

PHILADELPHIA:

CLAXTON, REMSEN & HAFFELFINGER,

624, 626 & 628 MARKET STREET.

1874.

Acknowledgment.

THROUGH the kindness, and liberality, of Professor *Joseph Henry*, Secretary of the Smithsonian Institution, we have been permitted to take electrotype copies of numerous wood-cuts, originally prepared for its own publications.* — In this way, we have been enabled, fully to illustrate the present work — an advantage which would, otherwise, have been unattainable, on account of the cost.

* The figures of *Unionidæ* were drawn from nature, and cut especially for the present purpose.

PREFACE.

THE idea has been long entertained, of collecting the *Natural Productions* of Chester County, with special reference to the elucidation of its *Prospective History*. With this object in view, THE CABINET OF NATURAL SCIENCE OF CHESTER COUNTY was organized, in the year 1826. From that time forward, several of its members engaged, with ardor, in the work. But it was soon discovered that the field for research was too ample — the labor too great — the laborers too few, for this *desideratum* to be speedily attained. Yet, much has been accomplished.

The indefatigable industry, and scientific acumen, of the late Doctor *William Darlington*, has left little more to be done, in *Phœnogamous and Filicoid Botany*, within our limits. His little FLORULA CÆSTRICA appeared in 1826. It was followed, eleven years after, by the more extended, and complete, FLORA CÆSTRICA. This was greatly enlarged, and improved, in a *third edition*, in 1853. It was also rendered more comprehensive, by the addition of the remaining *Anophyta* (the *Mosses* and *Liverworts*) furnished by *Thomas P. James;* and the *Thallophyta* (the *Lichens* and *Collemas*) by *Dr. E. Michener.*

ix

More recently, the last named gentleman has collected, and arranged, an extensive herbarium of *Hysterophyta* (he *Fungi*), a large proportion of which were in the county.

Thirty years ago, *Dr. E. Michener* collected, and preserved, specimens of most of the *Mammalia*, *Birds*, and *Reptiles*, known to inhabit Chester County. More recently, *Vincent Bernard* successfully occupied the same field. These important collections have mostly been placed in the Cabinet of Swarthmore College.

The proverbial richness of our county, in *rare* and *valuable minerals*, is amply sustained, and illustrated, by the splendid *Mineralogical Cabinets* of *William W. Jefferis*, and other *savans*.

Last, but not least; in 1853, *Dr. William D. Hartman* prepared "*A Descriptive Catalogue of the Terrestrial, and Fluviatile, Testaceous Mollusks, of Chester County*," containing fifty-nine species. The cost of printing prevented the publication at that time of more than "*A Classified Catalogue*" of the species.

More than a duplication of the number of species then known to inhabit the county, and the progressive improvements in classification and nomenclature, have rendered that *Catalogue* obsolete, and imperiously calls for one more in accordance with the advanced state of the science. Our wish, and desire, is *to supply this want.* And we entertain the hope that our effort will be received with indulgence; and, that it may serve to diffuse a taste for, and facilitate the study of, this interesting branch of our rich *Fauna Cestrica* among our young naturalists.

A few *extra-limital* species have been inserted, from the eastern slope of the mountain range, in our State; and which may yet occur within this county.

Our manuscript was prepared three years ago; and its publication delayed in order to provide the illustrations. The
rapid advance of the science, during this brief period, rendered it necessary that we should re-arrange, and re-write, a
large portion of it.

Copious *materiel* exists toward the remaining OLOGIES of
the Natural History of Chester County. Who will undertake
the task of making the necessary selections, and giving them
to the public?

CONCHOLOGIA CESTRICA.

THE MOLLUSCA.

THE term Mollusca, or Mollusks, is applied to a numerous division of the animal kingdom, composed, as the name implies, of soft-bodied, or gelatinous animals, without either an internal skeleton, or articulated appendages. They are either terrestrial (*Geophila*), fluviatile (*Limnophila*), or marine (*Thalasiophila*). Their respiration is either aerial (*pulmonary*), or aquatic (*branchial*). Some aquatic species are air-breathers, and may be termed *amphibious*. They are either naked, or testaceous; the latter being either univalve, or bivalve. They are either unisexual, or hermaphrodite.

Mollusks are, generally, more or less inclosed in a thick, dense skin (*the mantle*), which varies greatly in form and extent, in different species. It gives attachment to a highly complex muscular system of great irritability; which, when excited, pours out a copious secretion of mucus. In the testaceous, or shell-bearing species, the appropriate parts of the mantle secrete the matter which constitutes the shell, in all its variety of form and color.

The description, and classification, of those shells, so wonderfully varied in form, so exquisitely sculptured, so richly painted, ONCE CONSTITUTED CONCHOLOGY. It does not do so now. In the zoölogical, as in the

physical, sciences, the watchword is — *Onward.* The
form, the sculpture, the painting, of the shells have not
changed; but the inquiry has gone forth — Who are
the *artistic builders* of those beautiful forms? and what
is their *organic,* and *life history?* To answer these in-
quiries, they have been interrogated in every sea, on
every shore, on the land, and in the water, and sub-
jected to the severest microscopic scrutiny. The stores
of information, thus brought back, whether in its vast-
ness, or its minuteness, of detail, has been applied to
extend, and to improve, what was already known re-
specting the shells. This knowledge now constitutes
CONCHOLOGY.

But the end is not yet. The heaven-aspiring mind
of man will not rest here, satisfied. Having thus traced
the creature, so far as his finite powers can go, he will
be irresistibly led, by the spiritual aspirations of his
being,

> " To look through nature, up to nature's God,"

and to worship, and adore, that great, and good, Being
whose wisdom planned, and whose power created, all
those evidences of His omnipotence. To his mind,

> " Each moss, each shell,
> Each crawling insect, holds a rank,
> Important in the plans of Him
> Who fram'd their being." — *Stillingfleet.*

DIAGRAM OF THE HIGHER DIVISIONS REPRESENTED IN THIS CATALOGUE.

DIVISION.—MOLLUSCA.	CLASS.	ORDER.	SUB-ORDER.	FAMILIES.
	GASTEROPODA.	PULMONIFERA.	GEOPHILA, INOPERCULATA.	Philomycenidæ, Limacidæ, Helicidæ, Succinidæ, Pupadæ.
			LIMNOPHILA, INOPERCULATA.	Auriculidæ, Limnaeidæ.
			LIMNOPHILA, OPERCULATA.	Viviparidæ, Valvatidæ. Amnicolidæ, Strepomatidæ,
	ACEPHALA.	BRANCHIFERA.		Unionidæ, Corbiculidæ.

Class GASTEROPODA.

Animal, head distinct, tentacles four, sometimes two; eyes at the tip, or near the base, of the superior pair; body more or less enclosed by the mantle; beneath with a dense, elongated, flattened disk (the foot). In the shell-bearing species, the posterior part of the body is enclosed in the shell. The respiratory, and anal orifices, are placed near together on the right side of the neck. Bisexual, but requiring the union of two individuals for reciprocal fecundation.

Shell, mostly present, and spiral, as in the snails; sometimes only rudimentary, or absent, as in the slugs. When the shell is complete, the mantle covers the upper exterior parts of the animal, and forms a broad, flat ring, by its posterior margin, which closes the aperture of the shell when the animal is extended.

The animals comprised in this class are more highly organized, and their anatomy is more complex, than other molluscous animals. Besides the organs herein described, which pertain to their family, and generic, characters, they possess a very complicated and excitable **Muscular System**, many of the muscles being united to the skin. The ventral surface is covered with a thick muscular layer forming a long disc, which is termed the foot. The contraction of these fibres produce wrinkles, which succeed each other from behind forward, thereby enabling the animal to glide over solid surfaces, or on the water. In those possessing a turbinated shell, a large muscle arises from the columella, and after dividing, is spread over the sides of the body to be inserted into the foot; an arrangement which enables them rapidly to retract the body within the shell. Numerous other muscles within the body serve different uses in their economy.

The Nervous System, is composed of several ganglionic masses, connected by nerve filaments. The largest one is placed on the œsophagus, and is the functional representative of the brain.

The Circulation, is carried on by a heart consisting of an auricle, and ventricle, usually placed on the right side of the body, at the base of the branchia, in the aquatic; or in the pulmonary cavity of the terrestrial species. The blood is usually bluish-white; in the genus Planorbis it is red.

The Organs of Respiration, are either pulmonary, or branchial, and are for the most part lodged in a cavity on the right side of the body, just within the last whorl of the shell, the entrance to which is valvular, and may be seen opening and shutting at the will of the animal.

The Vent, is usually found just anterior to the pulmonary orifice. A third opening, near the base of the right tentacle, leads to the generative organs.

The Tentacles, (mostly four,) are either cylindrical, tubular, and retractile by inversion; or triangular, solid, conical, and contractile, with the head beneath the mantle.

The Eye, is very small, and placed on, or near, the base of the tentacles, and consists of the usual coats pertaining to that organ, together with a lens.

The Organ of Hearing, is composed of two auditive capsules, placed on the posterior surface of the large œsophageal ganglion (the brain). These capsules contain several otolites, or small spherical bodies, composed of carbonate of lime, immersed in a fluid, and possessing a rotary or vibratory motion, derived from the vibratile cilia within the capsules.

The situation of the Olfactory Organ, has not been satisfactorily ascertained. In the terrestrial species, it is supposed to reside in a *cul de sac*, between the lower lip and the front of the foot.

The Organ of Taste, is alike unknown; but that needful sense probably resides within the buccal cavity which contains the lingual ribbon.

The Oral Organs, are highly complicated — the lip' or palate is armed with a transverse arcuated maxillary plate or jaw, either entire, or consisting of several pieces; within this is an elongated strap of ligamentous or corneous matter, called the Tongue, or Lingual Ribbon; the surface of which is thickly set with Sharp Siliceous Teeth, beautifully arranged in rows, both transverse and longitudinal, with the points turned inward. The number of denticles varies from a few hundreds, to many thousands. The tongue is very retractile, by its own proper muscles; and is an ingestive, as well as a manducatory organ. When in use, it is quickly projected from the mouth in the form of a loop, and as rapidly withdrawn.

The Alimentary Canal, is long, and convoluted, and some species are provided with Salivary Glands.

The Kidneys, are sometimes present.

The Liver, always.

2 *   B

Order PULMONIFERA.

Terrestrial, or sub-aquatic animals, which respire free air, through an opening in or under the right side of the mantle. In some of the families of this order, the animal is furnished with a testaceous or horny plate, *the operculum*, attached to the posterior part of the foot. When the animal retires within the shell, the posterior half of the body is folded on the anterior, resembling in its action the shutting of a clasp-knife, after which, the head and body are withdrawn, leaving the operculum, which is situated on the posterior part of the foot, to close the aperture.

Sub-order GEOPHILA.

Animal, terrestrial, tentacles four, retractile by inversion, or contractile with the head beneath the mantle; upper pair long, cylindrical, ending in bulbs, which contain the eyes; lower ones short, or obsolete; *shell* mostly spiral.

INOPERCULATA.

The operculum wanting.

Family PHILOMYCENIDÆ.

The Slugs.

Mantle covering the entire upper surface of the animal. Shell entirely wanting.

DIAGRAM OF THE GENERA AND SPECIES OF THE FAMILY PHILO-
MYCENIDÆ.

FAMILY, { GENUS, TEBENNOPHORUS. { SPECIES, Carolinensis,
PHILOMYCENIDÆ. { SUB-GENUS, PALLIFERA. { dorsalis.

Obs. — The slugs, or naked snails, have many characters common to the true snails; both are nocturnal in their habits; both inhabit damp places under logs, or beneath loose bark, and decaying wood, or stones ; often among grass, and frequently in cellars and out-houses. But they do not hibernate; cold renders them torpid, but a little temporary warmth reanimates them; hence those infesting cellars, and green-houses, continue their depredations during the winter season. Slugs appear to be omnivorous, and may be found feeding, indiscrim-inately, on animal or vegetable substances. In gardens and orchards they are more herbivorous, often injuring tender plants and fruit.

Genus TEBENNOPHORUS, BINNEY, 1842.

Animal robust, subcylindrical, obtuse, or truncated posteriorly; foot somewhat expanded at the sides; generative orifice behind, and below the right superior tentacle.

T. Carolinensis, Bosc.
Limax Carolinensis, Bosc, Vers de Buff de Deters, 1830.

Tebennophorus Carolinensis. — [Binney & Bland.]
Fig. 1.

Grayish white, marbled with spots of black and brown, somewhat in three lines; foot whitish; respiratory orifice one-fourth inch behind the tentacle; lingual membrane with 115 rows of 113 teeth each, 56-1-56; buccal plate arcuate, with a slight denticulation on the concave margin; length 3 inches.

Jaw of T. Carolinensis. [B. & B.]
Fig. 2.

Station, among rotten wood, under loose bark, etc. Chester County; common.

Lingual Dentition of T. Carolinensis. — [B. & B.]
Fig. 3.

Sub-genus PALLIFERA, MORSE, 1864.

Animal, in external characters, resembling Tebennophorus; but differing in the form of the buccal plate, and dentition, of the lingual membrane.

P. dorsalis, BINN.

Pallifera dorsalis.
[B. & B.]
Fig. 4.

Philomycus dorsalis, Binn., Proc. Bost. Soct. Nat. Hist., 1841.

Cylindrical, attenuate, acute posteriorly, above obscurely elongate-rugose; ashy blue, with a black dorsal line; tentacles blackish;

Jaw of P. dorsalis. — [Morse.]
Fig. 5.

lingual membrane with 115 rows, of 113 teeth, each; 56-1-56; buccal plate arcuate, with costae strongly denticulating the concave margin; length ¾ of an inch.

Station, under logs, and decayed wood, near Philadelphia. (Tryon.)

Lingual Dentition of P. dorsalis. — [Morse.]
Fig. 6.

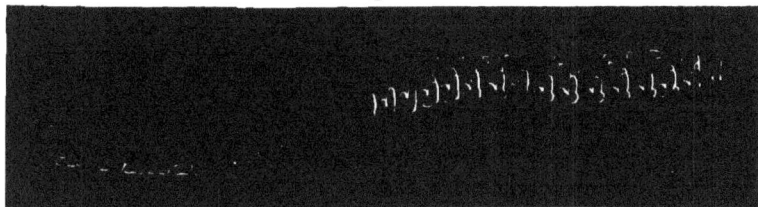

Family LIMACIDÆ.

Shell small, flat, rudimentary, and concealed within the mantle; which only covers the upper portion of the animal.

DIAGRAM OF THE GENERA, AND SPECIES, OF THE FAMILY LIMACIDÆ.

FAMILY.	GENUS.	SPECIES.
LIMACIDÆ.	LIMAX.	flavus, agrestis, campestris, maximus.

Genus LIMAX, LINN., 1740.

Animal, more or less elongated, tapering, acute, mantle occupying the anterior part of the body, wrinkled; respiratory orifice in the lower edge of the mantle. *Shell* very small, and entirely concealed.

L. flavus, LINN., Syst. Nat., Ed. x., i., 1758.

Limax flavus. — [B. & B.]
Fig. 7.

Animal, brownish-yellow, with rows of roundish white spots; head, neck, and upper tentacles, bluish; posteriorly acutely keeled; mantle large, gibbous, concentric-striate; pulmonary orifice cleft in the edge of the mantle; lingual membrane very broad, with 100 rows of 85 teeth each, 42-1-42; buccal

Jaw of L. flavus.
[B. & B.]
Fig. 8.

Lingual Dentition of L. flavus. — [B. & B.]
Fig. 9.

plate arcuate, ends square, with a projection on the concave margin; length 3 to 4 inches.

Station, in cellars, yards, and gardens. Chester County; frequent.

OBS. — An unwelcome foreigner from Europe, very common along our seaboard; especially in the cities, and adjacent to them.

L. agrestis, LINN., Syst. Nat., Ed. x., i., 1758.

Limax agrestis. — [B. & B.]
Fig. 10.

Animal, pallescent, rufescent, or nigrescent; maculated; tentacles blackish; mantle oval, gibbous, concentric-striate, one-third as long as the body; dorsal glands flattened, with the interspaces darker; respiratory orifice near the posterior, lateral, edge of the mantle, and bordered with white; length 1 to 2 inches.

Station, under decayed wood, boards, and stones, in cool places; everywhere common.

Obs. — This also is of European origin, and is now generally diffused.

L. campestris, Binn., Proc. Bost. Soct. Nat. Hist., 1841.

Animal, with·varying shades of amber, brown, and dusky; but unicolored, mantle ap-pressed, oval-oblong; dorsal glands elevated, elongated, with the inter-spaces unicolored; length near 1 inch.

L. campestris.—[B. & B.]
Fig. 11.

Station, similar to the last; Chester County.

Obs. — This species may be distinguished from L. agrestis by its smaller size, greater transparency, at all stages of growth, and in not secreting a milky mucus, when touched. It is probably indigenous.

L. maximus, Linn., Syst. Nat., Ed. x., i., 1758.

Limax maximus. — [W. G. B.]
Fig. 12.

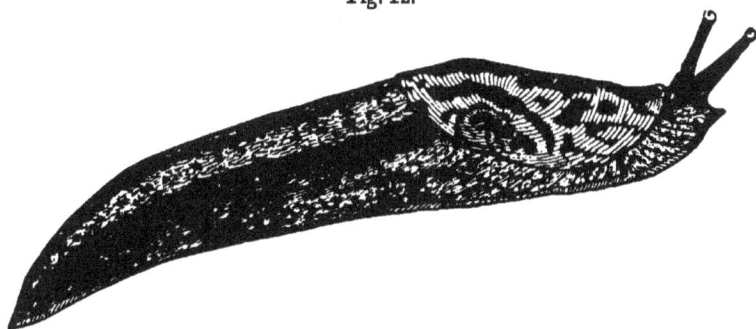

Animal, light brown, or ashen, with alternate rows of round spots, and uninterrupted stripes, of black, along the back, and sides; lighter on the sides; dirty white beneath; body elongate, with a well-marked dorsal carina, and covered with coarse, elongated, longitudinal, tubercles; mantle large, bluntly oval, with concentric tuberosities, and irregular black blotches; respiratory orifice on the posterior, lateral, edge; foot narrow; length 4 to 5 inches.

OBS. — An introduced species, from France; found in cellars, in Philadelphia. (Tryon.) And while this work is being printed, it has turned up in a cellar in West Chester, Chester County. Probably a direct importation from France, in connection with her wines. The figure, and description, has been reproduced from the admirable edition of Gould's Invertebrata of Massachusetts, by W. G. Binney, Esq.

Family HELICIDÆ.

The Snails.

Animal of Mesodon palliata. — [B. & B.]
Fig. 13.

Animal, very similar to the *Limacidæ;* except the posterior part of the body, which is spiral, and raised off the foot, to be inserted in the shell; the pulmonary orifice is in the collar, or margin, of the mantle, near the angle of the mouth of the shell; the anal opening is contiguous thereto. *Shell* discoidal, sub-orbicular, turbinate, or trochiform.

OBS. — The true snails are oviparous, and hermaphrodite; but require the union of two individuals, for reciprocal fecundation. They are both herbivorous and carnivorous. The eggs are usually deposited in places which they inhabit, to the number of fifty, or more. According to Dr. Binney, the depth is regulated by the distance the animal can penetrate the earth, while its shell remains above ground. They are globular, or roundly oval, sometimes slightly connected in bundles, or strings. The young appear in twenty days, or more, being influenced very much by moisture, and temperature; and are provided, at first, with a shell of about one and a half turns. Unable to subsist without moisture, the snails inhabit damp, and secluded, places, under stones, logs, and fallen leaves; often in pastures, lawns, and gardens. Unlike the slugs, they hibernate during winter, either in their summer retreats, or buried deeper in the earth. On the first occurrence of frost, they retire to their mural homes, close the aperture of the shell, with a thin transparent membrane, *the epiphragm;* and as the animal retreats, still farther, other epiphragms are successively formed.

DIAGRAM OF THE SUB-FAMILIES, SUB-GENERA, AND SPECIES, OF
THE FAMILY HELICIDÆ.

FAMILY.	SUB-FAMILIES.	SUB-GENERA.	SPECIES.
HELICIDÆ.	HELICINÆ.	MESODON.	albolabris, Sayii, Pennsylvanica, clausa, thyroides, bucculenta, " var. rufa, dentifera, palliata.
		TRIODOPSIS.	appressa, tridentata, fallax, introferens, " var. minor, inflecta.
		STENOTREMA.	monodon, hirsuta.
		ANGUISPIRA.	alternata.
	HELICELLINÆ.	MACROCYCLIS.	concava.
		ZONITES.	fuliginosus, laevigatus, inornatus, sub-planus, ligerus, demissus, gularis, suppressus, internus.
		HYALINA.	cellaria, indentata, electrina, arborea, hydrophila.
		PSEUDOHYALINA.	minuscula.
	VALLONINÆ.	VALLONIA.	minuta.
		PATULA.	striatella.
		STROBILA.	labyrinthica.
		HELICODISCUS.	lineatus.
	PUNCTINÆ.	PUNCTUM.	minutissimum.

3

Sub-genus MESODON, Rafinesque, 1831.

Animal of Mesodon albolabris. —[B. & B.]
Fig. 14.

Shell sub-globose, or orbiculate-depressed; umbilicus open, or covered; thin, finely striate, or decussate, sculptured; whorls 5 or 6, regular, aperture rounded, or lunate, sometimes with a parietal tooth; peristome broadly reflected, white, margin rarely unidentate.

M. albolabris, Say.
Helix albolabris, Say, Nich. Encyc., Amer. Ed., 1817.

M. albolabris. —[B. & B.]
Fig. 15.

Jaw of M. albolabris. —[Morse.]
Fig. 16.

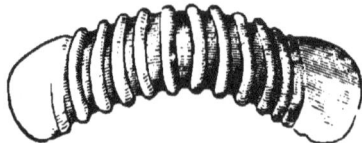

Shell large, convex, yellowish-brown, or russet; whorls 5 or 6, obliquely, fine-striate, with still finer revolving striæ, most distinct behind the reflected lip; lip broadly reflected, and flattened, closing the umbilicus when mature; lingual membrane of 123 rows, of 44-1-44 teeth; buccal plate heavy, arcuate, with ten large ribs, crenulating the margins of the plate. H. 15, W. 25, millimeters. Varies much in size.

Station, rocky woodlands. Chester County, frequent.

OBS. — Our largest Helix. In some localities it has a tooth on the pillar lip.

Lingual Dentition of M. albolabris. — [Morse.]

Fig. 17.

M. Sayii, BINN.

Helix diodonta, Say, Long's Exped., II., 1824. Sayii, Binn., Bost. Jour. Nat. Hist., III., 1840.

M. Sayii. — [B. & B.]
Fig. 18.

Shell depressed-orbicular; whorls 5-6, obliquely striate; aperture suborbicular; lip narrowly reflected, white, and with a tubercular tooth at base; pillar lip toothed; umbilicus narrow, deep; pale horn color. H. 15, W. 22, millimeters.

Station, in forests, Western Pennsylvania.

M. Pennsylvanica, GREEN.

Helix Pennsylvanica, Green, Cont. Maclur. Lyc., No. 1, 1827.

M. Pennsylvanica. — [B. & B.]
Fig. 19.

Shell turbinate, sub-globose, obliquely striate, with elevated ridges; whorls 6, convex; aperture sub-tri-

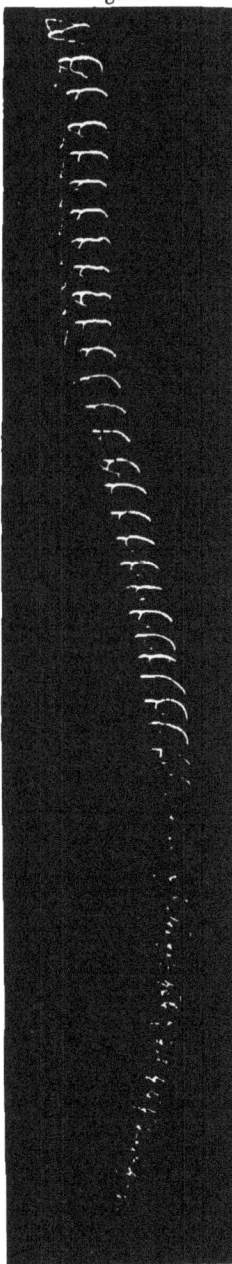

angular; lip narrowly reflected, white, slightly thickened, internally, at base; umbilicus closed, but indented.
Yellow horn color. H. 10, W. 18, mill.

Station, under logs, in woods. Western Pennsylvania.

M. clausa, SAY.

Helix clausa, Say, Jour. Acad. Nat. Sci. F. S., 1821.

M. clausa.
[B. & B.]
Fig. 20.

Shell sub-globose, light yellowish-brown;
whorls 5, finely striate; aperture rounded,
contracted; lip reflected, flat; umbilicus
narrow, and partly covered by the reflection
of the lip. H. 9, W. 16, mill.

Station, under logs, etc. New Garden,
Chester County; rare.

OBS. — This so nearly resembles some edentate forms
of the next species, as to occasion difficulty.

M. thyroides, SAY.

Helix thyroides, Say, Nich. Encyc., Amer. Ed., 1817.

M. thyroides. — [B. & B.]
Fig. 21.

Shell depressed - globular, pale yellowish - brown;
whorls 5, with fine transverse striæ; aperture rounded;
pillar lip with a prominent, oblique, white tooth; lip
widely reflected, sometimes with its face grooved; yellowish exterior; umbilicus small, partly covered. H.
12–14, W. 20–25, mill.

Station, in woods, gardens, and among decayed wood.
Chester County. Often confounded with the next.

M. bucculenta, GOULD.

Helix bucculenta, Gould, Proc. Bost. Soct. Nat. Hist., III., 1848.

M. bucculenta. — [B. & B.]
Fig. 22.

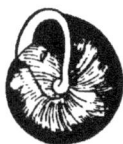

Shell sub-globose, finely striate, yellowish-white to pale brown; whorls 5, rounded; base convex; aperture rounded; the peristome forming nearly two-thirds of a circle, and broadly reflected, white, flesh colored behind; umbilicus partly covered; sometimes a small, white, tooth on the pillar, often edentate. H. 10, W. 15, mill.

Station, under logs in moist grounds, and meadows. Chester County; common.

M. bucculenta, var. Rufa, Michener, Amer. Jour. Conch., II., 1866.

Helix rufa, De Kay, Nat. Hist. N. Y., Part I., 1843.

Shell smaller and more delicate than the preceding; often strongly rufous, giving color to the reflected lip; the transverse striæ are strongly marked, and often reticulated, by microscopic revolving lines. H. 13, W. 18, mill.

Station, accompanying the last species. Chester County; common.

OBS. — W. G. Binney, speaking of *M. albolabris*, says, "Helix rufa, De Kay, appears to be the young of this species." Again, under *H. thyroides*, he says: "One [variety] from Germantown, Pa., is very small, measuring only 15 millimeters in diameter. It is globose, shining, sometimes imperforate, and generally without a parietal tooth. It is impossible to distinguish it from some forms of *H. bucculenta.*" The impossibility arises from the fact that it is, really, one of those forms.

M. dentifera, BINN.

M. dentifera.
[B. & B.]
Fig. 23.

Helix dentifera, Binn., Bost. Jour. Nat. Hist., I., 1840.

Shell depressed, spire sub-convex; base well rounded; whorls 5, fine striate; aperture wide; lip broadly reflected, and covering the umbilicus; parietal tooth prominent; color yellow horn. H. 10, W. 23, mill.

Station, mountain forests, Pennsylvania.

M. palliata, SAY.
Helix palliata, Say, Jour. Phil. Acad. F. S., II., 1821.

M. palliata.
[B. & B.]
Fig. 24.

Shell depressed, brown, rough or hispid; whorls 5, flattened above, striate; aperture three-lobed, contracted; lip reflected, white, often edged with brown, with two intermarginal teeth; pillar lip with a long, white, curved, tooth, originating at the umbilicus, which is covered. H. 9, W. 35, mill.

Station, mountain forests. Western Pennsylvania.

Sub-genus TRIODOPSIS, RAFINESQUE, 1819.

Shell orbiculate-depressed, or sub-globose; obliquely striate; whorls 5–7, the last sub-deflected; umbilicate, or closed; aperture sinuate, sub-triangular; peristome with a broad reflected margin; sides of the aperture, armed with strong teeth.

T. appressa, SAY.

Helix appressa, Say, Jour. Phil. Acad. F. S., II., 1821.

Shell much depressed, pale yellowish; whorls 5, often sub-angulated, with fine transverse striæ; aperture contracted; the white, reflected, lip, close appressed at base, and covering the umbilicus; the inner margin of the lip, mostly with one or two teeth; an oblique, compressed, white, tooth on the pillar lip; lingual membrane with 105 rows, of 40-1-40 teeth, each; buccal plate strongly arcuate, the ribs, on the anterior surface, denticulating both margins. H. 7, W. 17, mill.

Lingual Dentition of T. appressa. — [B. & B.]
Fig. 27.

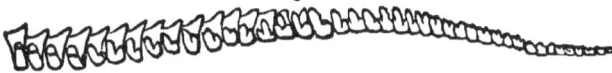

Station, under flat stones. On the Lehigh River, Pennsylvania.

T. tridentata, SAY.

Helix tridentata, Say, Nich. Encyc., Amer. Ed., 1817.

Shell very much depressed, light horn color, to chestnut; whorls 5, obliquely acute-striate; aperture trilobate, contracted by a groove behind the lip; lip reflected, white, with two acute, projecting, teeth, on its inner edge; pillar lip with a strong, oblique, tooth in the aperture; striæ converging into the deep, open, umbilicus. H. 8, W. 15, mill.; very variable.

Station, in fields bordering woods. Chester County.

T. fallax, SAY.

Helix fallax, Say, Jour. Phil. Acad. F. S., V., 1825.

Similar in outline; but differs from T. tridentata, in being much smaller; the spire is more elevated; the teeth are larger; and the upper one, of the lip teeth, is strongly inflected. H. 7½, W. 13, mill.

Station, in woods, under leaves and decayed wood. Chester County; common.

T. introferens, BLAND.

Helix introferens, Bland, Ann. N. Y. Lyc. Vij., 1860.

Shell *narrowly* umbilicate, depressed-globose, thin, costate-striate, pale; whorls 6, the last scarcely deflected, but deeply constricted at the aperture, with *two exterior pits;* periphery subangular; convex beneath, with *a spiral groove within the umbilicus;* aperture oblique, lunate, with a strong, arcuate, parietal tooth; peristome reflected, thickened within; an obtuse, inflected, tooth, within the right margin; at base, a submarginal, lamelliform, tooth, with a transverse tubercle in the centre; the basal lamella is continued within the aperture, where it forms a strong, white, tubercle. H. 7, W. 15, mill.

Station, in woods, under loose bark, and moist decayed leaves. Chester County.

T. introferens, var. Minor, BLAND.

Helix introferens, var. Minor, Bland, Ann. N. Y. Lyc. Vij., 1860.

Shell smaller; whorls 5. H. 6, W. 11, mill.

Station, in woods, gardens, and among rubbish. Chester County; abundant.

T. inflecta, SAY.

Helix inflecta, Say, Jour. Phil. Acad. F. S., II., 1821.

Shell convex, brownish horn color, with fine, hair-like, projections; whorls 5, with very minute, transverse, striæ; suture not much impressed; aperture three lobed, very much contracted, by a deep groove behind the lip; lip white, narrow, reflected over the umbilicus, with two acute teeth on the inner margin; the superior one inflected; the other, half way between it and the base, and separated by a profound sinus; parietal tooth long, arcuated; umbilicus covered, impressed. H. 7, W. 12, mill.

T. inflecta.— [B. & B.]
Fig. 31.

Station, mountains, Pennsylvania.

Sub-genus STENOTREMA, RAFINESQUE, 1819.

Shell lenticular, or depressed globose, mostly pilose; whorls 4½–6, the last gibbous; base tumid; aperture long, and narrow; basal lips white; the superior narrowly reflected; the inferior similar, or presenting a notch, near the middle; no umbilicus.

S. monodon, RACKET.

Helix monodon, Racket, Linn. Trans., XIII., 1822.

S. monodon.
[B. & B.]
Fig. 32.

Shell sub-globose, depressed above, pale chestnut, minutely hairy; whorls 5–6; aperture arcuate, contracted, by a deep groove, behind the lip, which is narrowly reflected, and reaches to the base, partly, or wholly, closing the umbilicus; parietal tooth elongated; lingual membrane with 100 rows, of 28-1-28, teeth, each; buccal

Jaw of S. monodon.
[Morse.]
Fig. 33.

C

plate, slightly arcuate; anterior face, with broad ribs, denticulating each margin. H. 3½–4½, W. 7–10, mill.

Lingual Dentition of S. monodon. — [Morse.]
Fig. 34.

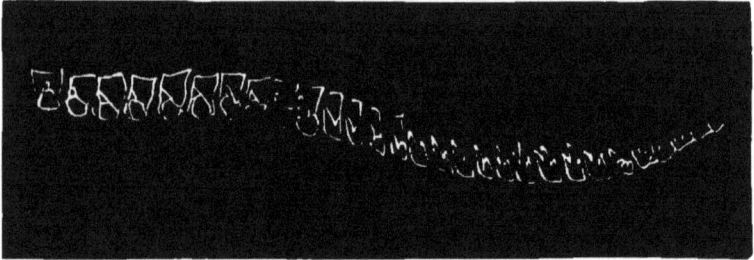

Station, under decayed bark and stones. Chester County; rare.

S. hirsuta, SAY.

Helix hirsuta, Say, Jour. Phil. Acad. F. S., I., 1817.

S. hirsuta.
[B. & B.]
Fig. 35.

Shell sub-globose, pale brown, with sharp, rigid, hairs, arranged in oblique lines; whorls 5; aperture very narrow, almost closed, by an elongated, lamelliform, tooth, on the pillar lip, reaching from the centre of the base to within the junction of the lips, above, and covering the umbilicus; outer lip narrowly reflected, grooved, with a notch near the middle of its inner margin; base very convex. H. 4, W. 6, mill.

Station, under loose bark of logs. Chester County; abundant.

OBS. — In some localities, it nearly doubles the dimensions here given.

Sub-genus ANGUISPIRA, MORSE, 1864.

Shell heavy, large, depressed-turbinate, solid, costate-striate, banded with colored stripes; umbilicus moderate; aperture simple, without teeth.

A. alternata, SAY.

Helix alternata, Say, Nich. Encyc., Amer. Ed., 1816.

A. alternata. — [B. & B.]
Fig. 36.

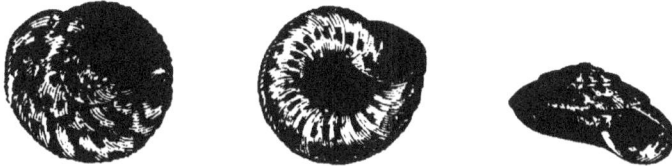

Shell orbiculate-depressed, obliquely, and closely, ribbed-striate; whorls 6, varied with oblique, rufous bars, and spots; convex beneath, often angular at the periphery; aperture oblique; umbilicus large; base

Jaw of A. alternata.
[Morse.]
Fig. 37.

paler, and less striate; lingual membrane, with 121 waving rows, of 34-1-34, teeth, each; buccal plate stout,

Lingual Dentition of A. alternata. — [Morse.]
Fig. 38.

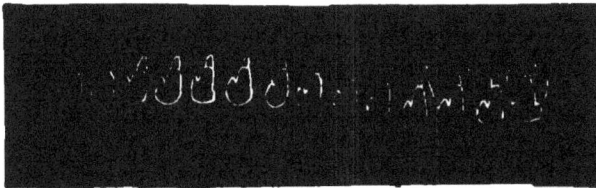

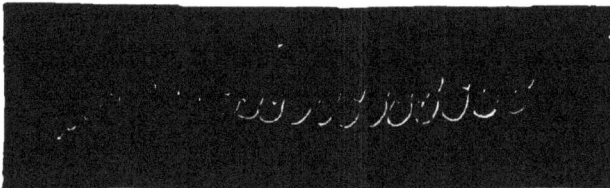

of uniform width, and deeply striate, longitudinally, and transverse. H. 8–10, W. 20–25, mill.

Station, under wet logs, and boards, along streams. Chester County; common.

Sub-genus MACROCYCLIS, BECK, 1837.

Animal of Macrocyclis concava. — [B. & B.]
Fig 39.

Shell moderate, thin, wide, umbilicate, depressed, striate, unicolored; whorls 4–5, the last wide, depressed, and deflected; aperture oblique, ovate; peristome slightly thickened, and reflected at base.

M. concava, SAY.
Helix concava, Say, Jour. Phil. Acad. F. S., II., 1821.

M. concava.
[B. & B.]
Fig. 40.

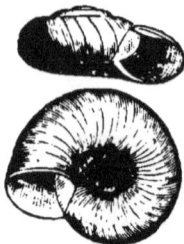

Shell convex-discoid, pale horn color, sometimes with a tinge of green; whorls 5, substriate, the last one flattened near the mouth, rounded, beneath; umbilicus wide, and deep; lip sub-reflexed, at base, and its extremities united by a columellar deposit; lingual membrane with — rows, of 23-1-23, teeth, each; buccal plate crescentic, anterior surface striated, concave margin, with a slight projection. H. 7, W. 16, mill.

Lingual Dentition of M. concava. — [B. & B.]
Fig. 42.

Jaw of M. concava.
[B. & B.]
Fig. 41.

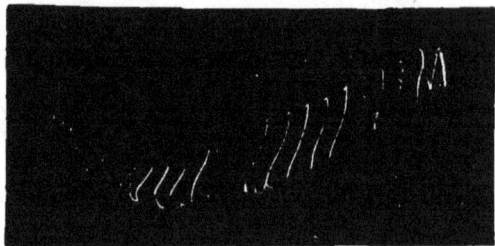

Station, under stones, logs, etc. Chester County; not common.

Sub-genus ZONITES, MONTFORT, 1810.

Animal of Zonites fuliginosus. —[B. & B.]
Fig. 43.

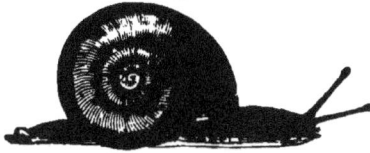

Shell orbiculate, convex, or discoidal; whorls 6–7, gradually increasing, striated above, smooth beneath; aperture lunate, oblique; peristome acute, and slightly thickened, within.

Z. fuliginosus, GRIFFITH.

Helix fuliginosa, Griff. (in litt.), C. B. Adams' Amer. Jour. Science, Vol. 40, p. 273.

Z. fuliginosus.
[B. & B.]
Fig. 44.

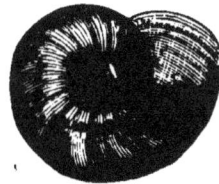

Shell rather large, thin, chestnut-brown, orbicular, depressed; whorls 4½, increasing rapidly; aperture lunate-ovate, very oblique; lip thin, acute, slightly thickened within, by a testaceous deposit; extremities of the peristome approximate; umbilicus moderate, deep; lingual membrane very broad, with 87 rows, of 64-1-64, teeth; buccal plate arcuate, margin simple, with a blunt, median projection. H. 13, W. 25, mill.

Jaw of Z. fuliginosus.
[B. & B.]
Fig. 45.

Station, in woods. Lancaster County, Pennsylvania.

Lingual Dentition of Z. fuliginosus. —[B. & B.]
Fig. 46.

Z. lævigatus, PFR.

Helix lævigata, Pfr., Mon. Hel. Viv., I., 1848.

Z. *lævigatus.*
[B. & B.]
Fig. 47.

Shell globosely depressed, thin, greenish horn color; whorls 5, closely striate above, smooth, and shining, beneath; last whorl rapidly expanding; aperture rounded-lunular; lip simple, much thickened, within the base, and slightly reflected, around a moderate umbilicus. H. 9, W. 18, mill.

Station, mountains. Sunbury, Pennsylvania.

Z. inornatus, SAY.

Z. *inornatus.*
[B. & B.]
Fig. 48.

Helix inornata, Say, Jour. Acad. Phila. F. S., II., 1821.

Shell much depressed, smooth, shining, pale yellow; whorls 5, flattened above; aperture transverse; lip thin, acute, with an internal, white, deposit; umbilicus small, deep. H. 6, W. 16, mill.

Station, in woods. Lancaster County, Pennsylvania.

Z. subplanus, BINN.

Z. *subplanus.*
[B. & B.]
Fig. 49.

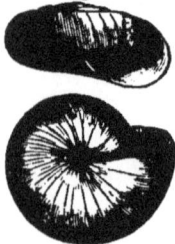

Helix subplana, Binn., Bost. Jour. Nat. Hist., IV., 1842.

Shell small, planulate, above, and beneath, greenish-brown, shining; whorls 5½, fine striate, near the apex; aperture transverse; lip simple; umbilicus small, round, and deep. H. 6, W. 18, mill.

Station, mountains. Western Pennsylvania.

Z. ligerus, SAY.

Helix ligera, Say, Jour. Acad. Phila., F. S., II., 1821.

Shell sub-globose, obtuse, brownish-yellow above, paler beneath; whorls 6–7, oblique-striate; aperture lunate; lip acute, with a strong deposit within the margin, near its base; umbilicus very small, or none. H. 6–8, W. 11–15, mill.

Z. ligerus.
[B. & B.]
Fig. 50.

Station, in woods. Chester County; not common.

Z. demissus, BINN.

Helix demissa, Binn., Bost. Jour. Nat. Hist., IV., 1843.

Shell depressed, yellowish, thickly striate; base somewhat flattened, smooth; aperture sub-transverse, compressed-lunate; umbilicus very small; lip acute, thickened within the margin. H. 6, W. 10–12, mill.

Z. demissus.
[B. & B.]
Fig. 51.

Station, in mountainous districts. Western Pennsylvania.

Z. gularis, SAY.

Helix gularis, Say, Jour. Acad. Phila. F. S., II., 1822.

Shell depressed-conical, shining, dusky-yellow; whorls 7–8, very minute at the apex; aperture transverse; lip thickened within, at base, with one, or more, elongated, lamelliform, teeth; one at the base, the other more central; umbilicus covered, or very small. H. 5–7, W. 7–9, mill.

Z. gularis.
[B. & B.]
Fig. 52.

Station, under wet logs, and decayed leaves. Lancaster County, Pennsylvania.

Z. suppressus, SAY.

Z. suppressus.
[B. & B.]
Fig. 53.

Helix suppressa, Say, New Harm. Diss., II., 1829.

Shell small, yellowish-brown, thin, polished; whorls 6, slowly increasing, fine striate above, smooth beneath; lip thin, callous within the base; within the outer lip are one, or two, elongated, lamelliform teeth; umbilicus very small, or absent. H. 4, W. 6, mill.

Station, in fields, and woods, under stones, timber, etc. Chester County; abundant.

Z. internus, SAY.

Helix interna, Say, Jour. Acad. Phil. F. S., II., 1822.

Z. internus.
[B. & B.]
Fig. 54.

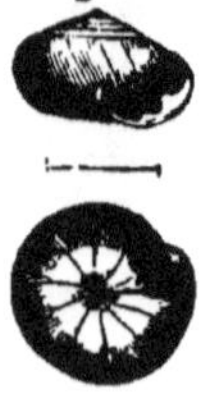

Shell convex-orbicular, reddish brown, above ribbed, beneath smooth; whorls 8, narrow, suture deep; periphery sub-angular; aperture transverse, narrow, thickened within, at the base, with two short lamellæ, near the basal margin; lip reaching to the axis; umbilicus nearly, or sometimes quite, closed. H. 4, W. 6, mill.

Station, similar to the preceding. Western Pennsylvania.

Sub-genus HYALINA (FERR.), GRAY, 1840.

Animal of Hyalina cellaria. — [B. & B.]
Fig. 55.

Shell mostly small, umbilicate, or closed, and indented; depressed, vitreous, shining; whorls 5–6, increasing regularly; peristome thin, acute; aperture round-lunate, not angled or depressed.

H. cellaria, MÜLL.

Helix cellaria, Müll., Hist. Verm., II., 1773.

H. cellaria.—[B. & B.]
Fig. 56.

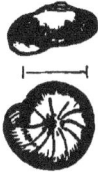

Jaw of H. cellaria, young and old.
[Morse.]
Fig. 57.

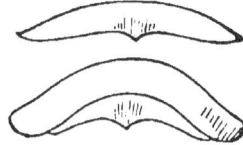

Shell depressed, polished, light greenish above, pale beneath; whorls 5, flattened; aperture oblique, subcircular; umbilicus small; lip thickened within, and subreflected near its base; lingual membrane with 38 curving

Lingual Dentition of H. cellaria.—[Morse.]
Fig. 58.

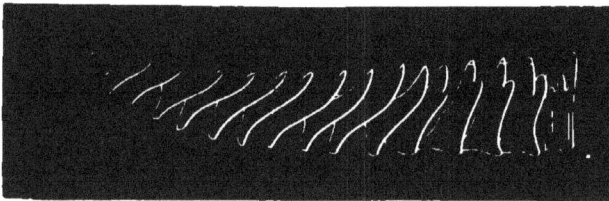

rows of 17-1-17 teeth, each; buccal plate strongly arcuate, with a median projection. H. 7, W. 13, mill.

Station, in cellars and gardens, in and near maritime cities. Chester County; Philadelphia.

OBS. — Introduced from England.

H. indentata, SAY.

Helix indentata, Say, Jour. Acad. Phila. F. S., II., 1822.

H. indentata.
[B. & B.]
Fig. 59.

Shell small, thin, depressed, pellucid, pale, highly polished; whorls 4, rapidly enlarging, with distant radiating, impressed, lines, which extend to the indented centre of the imperforate base; aperture expanded; peristome thin, acute, reaching to the basal centre. H. 2½, W. 5, mill.

4 *

Station, under logs and loose bark. Chester County; frequent.

H. electrina, GOULD.
Helix electrina, Gould, Invert., Mass., 1841.

H. electrina.
[B. & B.]
Fig. 60.

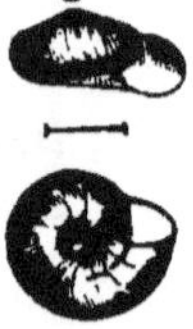

Shell very thin, pale; whorls $3\frac{1}{2}$, the last rapidly enlarging; spire depressed; aperture rounded; peristome simple, acute; lip not flexuous; umbilicus very narrow, and deep. H. 2, W. 4, mill.

Station, in company with the preceding. Chester County; rare.

H. arborea, SAY.
Helix arboreus, Say, Nich. Encyc., Amer. Ed., 1817.

H. arborea.
[B. & B.]
Fig. 61.

Shell much depressed, thin, shining, amber-colored; whorls $4\frac{1}{2}$, increasing regularly; umbilicus moderate, deep; lip slightly flexuous; lines of growth indistinct. H. $2\frac{1}{2}$, W. 5, mill.

Station, under decayed wood, and rubbish. Chester County; abundant.

H. hydrophila, INGALLS.
Helix hydrophila, Ingalls, in collection unpublished.

H. hydrophila.
[B. & B.]
Fig. 62.

Shell depressed-conical, thin, amber-colored; whorls 4 to $4\frac{1}{2}$; suture well impressed; umbilicus deep, crateriform; aperture lunate, sub-deflected; lip thin, sub-angular. H. 3, W. 6, mill.

Station, under logs, and boards, near the water. Schuylkill tide marsh, Philadelphia.

OBS. — We prefer to keep this distinct from H. nitida, Müll., of Europe.

Sub-genus PSEUDOHYALINA, MORSE, 1864.

Shell minute, discoidal, slightly convex above, uni-colored; closely striate, or ribbed; umbilicus large.

P. minuscula, BINNEY.

Helix minuscula, Binn., Boston Jour. Nat. Hist., III., 1840.

P. minuscula.
[B. & B.]
Fig. 63.

Shell slightly convex, whitish; whorls 4, increasing slowly in diameter; suture deep; umbilicus large; aperture rounded, spreading; lip thin, acute; columella with a thin callus; lingual membrane with 52 rows of 12-1-12 teeth, each; buccal plate slightly arcuate, long, and narrow; anterior surface with faint central, longitudinal striæ, slightly raised, on the centre of the cutting edge. H. 1, W. 2–2½, mill.

Jaw of P. minuscula.
[Morse.]
Fig 64.

Lingual Dentition of P. minuscula. — [Morse.]
Fig. 65.

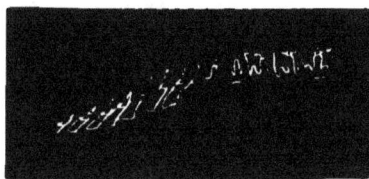

Station, in grass-fields, under sticks and stones. New Garden, Chester County; not common.

Sub-genus VALLONIA, RISSO, 1826.

Shell umbilicate, depressed; whorls 3½–4; aperture oblique, semicircular; peristome white; lip reflected, the margins approximate.

V. minuta, SAY.

Helix minuta, Say, Jour. Acad. Phila. F. S., I., 1817.

V. minuta, enlarged. — [B. & B.]
Fig. 66.

Jaw of V. minuta. — [Morse.]
Fig. 67.

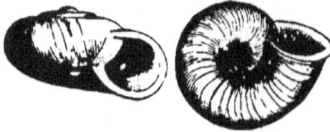

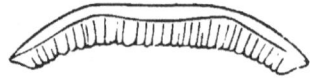

Shell minute, slightly convex, opaque white; whorls 4, fine striate; aperture orbicular, dilated; lip thick, broadly reflected, white, forming almost a circle; umbilicus wide, and deep; lingual membrane with 75 rows of 11-1-11 teeth, each; buccal plate wide, narrow; ends slightly bent, and longitudinally striate, the striæ extending to the cutting edge, producing minute notches. H. 1½, W. 2½, mill.

Lingual Dentition of V. minuta. — [Morse.]
Fig. 68.

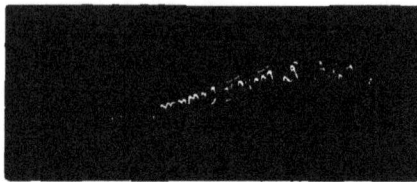

Station, among grass, under boards and rubbish. Chester County. Everywhere abundant.

Obs. — Until recently, a majority of writers on conchology, considered Say's species identical with the European *V. pulchella*, Müll. Prof. E. S. Morse, in his admirable paper on " *The Terrestrial Pulmonifera of Maine*," has pointed out the difference between the two species. These are shown in the accompanying outline figures, copied from the above work — Fig. 69, V. minuta, Say; and Fig. 70, V. pulchella, Müll. By a comparison of these figures it will be evident that our

shell is more depressed; the whorls are smaller; the
aperture is wider, and less round; and the angle of

V. minuta. — [Morse.]
Fig. 69.

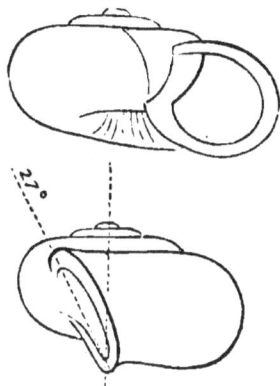

V. pulchella. — [Morse.]
Fig. 70.

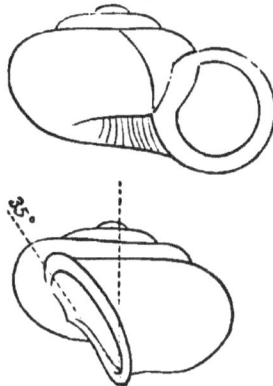

aperture, is 27°; while in V. pulchella it is 35°. They
also differ in the dentition of the lingual membrane;
and the periostraca. Both species agree in presenting
varieties which are strongly, and transversely, costate.

Sub-genus PATULA, HELDWIG, 1837.

Shell umbilicate, turbinate, or depressed-discoid; rugose or costate-striate; whorls 4–5, sub-equal; aperture
lunate-rotundate; peristome simple, acute, straight.

P. striatella, ANTHONY.

Helix striatella, Anthony, Bost. Jour. Nat. Hist., III.,
1840.

P. striatella. — [B. & B.]
Fig. 71.

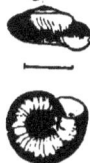

Jaw of P. striatella. — [Morse.]
Fig. 72.

Shell depressed-convex, reddish horn color, or brown;

whorls 3–4, strongly oblique-striate; suture lightly impressed; aperture rounded, or transverse; umbilicus

Lingual Dentition of P. striatella.—[Morse.]
Fig. 73.

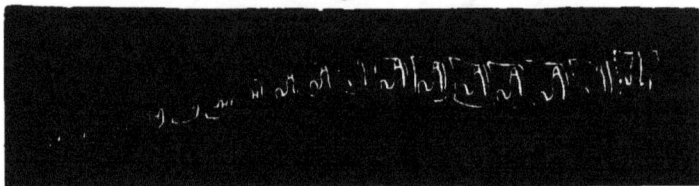

expanded, shallow; lingual membrane with 100 rows of 16-1-16 teeth, each; buccal plate arcuate, surface with converging striæ, and irregular notches on the concave margin.   H. 2½, W. 5, mill.

Station, under decayed bark, and leaves.   New Garden, Chester County; frequent.

Sub-genus STROBILA, MORSE, 1864.

Animal of Strobila labyrinthica.—[Morse.]
Fig. 74.

Shell very small, turbinate, with revolving laminated teeth, within both the columellar, and basal, lips; whorls numerous, strongly striate-costate; lip simple.

S. labyrinthica, SAY.

Helix labyrinthica, Say, Jour. Acad. Phila. F. S., I., 1817.

Shell obtuse-conic, reddish brown; whorls 6, heavily ribbed, above, more smooth beneath; lip thickened, somewhat reflected; base flattened; umbilicus small; aperture narrow-lunate, with three revolving lamina, on the parietal wall — one at the base; and two, far within the outer lip, near the base; lingual membrane with 78 rows of 13-1-13 teeth, each; buccal plate slightly arcuate, minutely notched on the cutting edge. H. 2½, W. 2½, mill.

S. labyrinthica, enlarged.
[B. & B.]
Fig. 75.

Jaw of S. labyrinthica.
[Morse.]
Fig. 76.

Lingual Dentition of S. labyrinthica. — [Morse.]
Fig. 77.

Station, under the bark of decayed logs. Chester County; frequent.

Sub-genus HELICODISCUS, Morse, 1864.

Animal of Helicodiscus lineata, enlarged. — [Morse.]
Fig. 78.

Shell minute, flat; whorls equally visible, above and below, with revolving striæ; unicolored, whitish; one, or more, lamellar teeth, within the outer lip.

H. lineata, SAY.

H. lineata, en-larged.—[B. & B.]
Fig. 79.

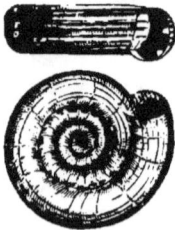

Helix lineata, Say, Jour. Acad. Phila. F. S., I., 1817.

Shell small, discoid, pale greenish yellow; whorls 4, planulate above, concave beneath, with about fifteen revolving lines; aperture narrow-lunate, two pairs of teeth, within the outer lip; one pair visible, the other deeper; lingual membrane with 77 curving rows of 12-1-12 teeth, each; buccal plate narrow, crescentic, anterior surface, with converging striæ, cutting margin smooth, with a median projection. H. 1, W. 3, mill.

Jaw of H. lineata. — [Morse.]
Fig. 80.

Lingual Dentition of H. lineata. — [Morse.]
Fig. 81.

Station, in cool wet places, under stones. Chester County; frequent.

Sub-genus PUNCTUM, MORSE, 1864.

Shell exceedingly minute, discoid, umbilicate; aperture rounded; peristome thin, acute.

P. minutissimum, LEA.

Helix minutissima, Lea, Trans. Amer. Phil. Soct., IX., 1841.

Shell very minute, depressed-turbinate, above, convex

below; pale fuscous, minutely striate; whorls 4; aperture transversely-lunate; umbilicus wide; lingual mem-

P. *minutissimum.*
[B. & B.]
Fig. 82.

Jaw of P. minutissimum.
[Morse.]
Fig. 83.

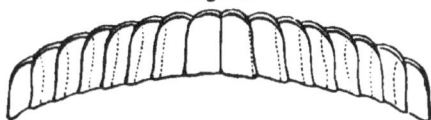

brane with 54 arched rows of 13-1-13 teeth, each; buccal plate consisting of sixteen, corneous lamina, partially

Lingual Dentition of P. minutissimum. — [Morse.]
Fig. 84.

overlapping, and recurved, on their cutting edge. H. 1, W. 1½, mill.

Station, among fallen leaves. Western Pennsylvania.

Family SUCCINIDÆ.

Shell oblique-ovate, imperforate, thin, pellucid, unicolored; spire very small; body whorl large, inflated; aperture large, oval, or ovate; peristome simple, acute; animal resembling that of Helix; tentacles short, conoid.

OBS. — These animals mostly affect low grounds, along the margin of streams, or where it is subject to overflow; while others are found only on high ground, remote from water. When supplied with abundant food, and moisture, they seem almost too large to enter fully into their shells; when these fail them, and on the approach of cold weather, this difficulty ceases. In organization, they are very much like the common snail, and their general habits are also very similar.

5   D

Genus SUCCINEA, Draparnand, 1801.

Animal of Succinea.
Fig. 85.

Generic characters as above.

S. ovalis, Gould (non Say), Invert., Mass., 1841.

Shell ovate-conic, very thin, shining, pale amber;
spire acute; whorls 3, the last elongated; aperture produced, broadly rounded below, more than three-fourths of the total length; lingual membrane with 80 rows of 40-1-40 teeth, each; buccal plate strongly arcuate, with three small folds on the cutting edge.   H. 12, W. 6, mill.

S. ovalis.
[B. & B.]
Fig. 86.

Jaw of S. ovalis.
[Morse.]
Fig. 87.

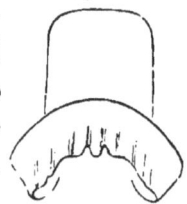

Lingual Dentition of S. ovalis. — [Morse.]
Fig. 88.

Station, on plants, and among wet chips, in autumn. Chester County; frequent.

S. obliqua, Say, Long's Exped., II., 1824.

S. obliqua.
[B. & B.]
Fig. 89.

Shell ovate, thin, shining, striate; whorls 3, last whorl ovate; aperture oval, both sides equally curved; nearly three-fourths of the total length; yellowish green. H. 20–25, W. 12–13, mill.

Station, on plants growing in and near water.   Chester County; frequent.

S. avara, SAY, Long's Exped., II., 1822.

Shell sub-oval, pale reddish-yellow; whorls 3, rounded, with a deep suture; aperture sub-ovate, *S. avara, enlarged.*
two-thirds the whole length. H. 6, W. [B. & B.]
3½, mill. Fig. 90.

Station, in damp pastures, lawns, etc. Chester County; common.

Family PUPADÆ.

Shell mostly small, cylindrical, ovate, or ovate-conical, and elongated; whorls numerous, the last not expanded; aperture sub-circular, simple, or armed with denticles; lip simple, or reflected; umbilicate. Animal twice as long, as broad; broad, and square, in front; head separated from the foot beneath, by a transverse groove; head transverse; tentacles four, the upper pair, occuliferous.

OBS. — These pygmy mollusks are so minute as to elude observation, unless specially sought for, in the stations they inhabit. Some species are found in woods, under the bark of decayed timber, while others occur more abundantly under stones, boards, or chips, near the margin of streams. Several species of *Vertigo* are found abundantly, in grass fields, and lawns, or among moss. In the latter stations, they may be readily captured during the summer and early fall months by placing a board on the wet grass in the evening; to the under surface of which they will be found closely adhering on the next morning. Like most of their class, they are vegetable feeders. In winter, they bury themselves in the ground, or beneath decayed leaves.

DIAGRAM OF THE SUB-FAMILIES, GENERA, AND SPECIES, OF THE
FAMILY PUPADÆ.

FAMILY.	SUB-FAMILIES.	GENERA.	SPECIES.
	ZUINÆ.	ZUA.	lubricoidea.
		ACICULA.	acicula.
PUPADÆ.	PUPINÆ.	LEUCOCHILA.	fallax, marginata, pentodon, armifera, contracta, corticaria.
	VERTIGINININÆ.	VERTIGO.	ovata, Gouldii, milium, decora.

Sub-genus ZUA, LEACH, 1820.

Animal of Zua, enlarged. — [Reeve.]
Fig. 91.

Shell small, elongated, cylindrical, or sub-conic;
whorls numerous; apex sub-obtuse; last whorl elongated, half as long as the shell; aperture ovate; lip
thin, sub-effuse at base; no umbilicus.

Z. lubricoidea, STIMPSON.
Bulimus lubricoides, Stimp., Test. Moll. N. E., 1851.
Bulimus lubricus, Say, and others.

Z. lubricoidea. — [B. & B.]   *Jaw of Z. lubricoidea.* — [Morse.]
Fig. 92.   Fig. 93.

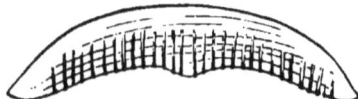

Shell oblong-oval, sub-acute, shining, pale brown;
whorls about 6, slightly rounded; aperture lateral, oval;

lip simple, with the margin thickened, and often rufous; lingual membrane, with 90 rows of 43 teeth, each; buccal plate slightly arcuate, with conspicuous longi-

Lingual Dentition of Z. lubricoidea.—[Morse.]
Fig. 94.

tudinal striæ; cutting edge with an obtuse beak in the centre. H. 5, W. 2, mill.

Station, under the bark of decayed timber, and moist decayed leaves. Chester County; frequent.

OBS.— This shell was, for a long time, supposed to be identical with the English species, *Zua lubrica*, Müller. To this opinion, the late Dr. Stimpson was the first to demur, on the score of its wide diffusion over this Continent; and a further examination, by Prof. E. S. Morse, tends to confirm the suggestion of Dr. Stimpson, as to its specific distinction. Fig. 95, *a*, represents an enlarged view of the central and lateral denticle of Zua lubricoidea, Stimp.; and Fig. 95, *b*, the same denticles

a, Dentition of Z. lubricoidea. b, Dentition of Z. lubrica,—[Morse.]
Fig. 95.

of Zua lubrica, Müll. Our shell also differs in other respects from its foreign analogue.

5 *

Sub-genus ACICULA, Risso, 1826.

Animal of Acicula, enlarged. — [Reeve.]

Fig. 96.

Shell slender, elongate, turreted, imperforate, thin, polished; whorls 6 or 7, the last rounded at base; columella slightly twisted, truncated at base, aperture oblong, peristome simple, acute.

A. acicula, Müller.

A.acicula,
enlarged.
[B. & B.]
Fig. 97.

Buccinum acicula, Müll., Hist. Verm., II., 1774.

Shell very delicate, transparent, bluish-white, cylindrical; whorls 7, the last one more than half the whole length; aperture oblong-ovate. H. 5, W. 1½, mill.

Station, green-houses, and nurseries. New Garden, Chester County; rare.

Obs. — An exotic species, sometimes imported with nursery and green-house plants.

Sub-genus LEUCOCHILA, Albers, et Martiens, 1860.

Animal of Leucochila pentodon.

Fig. 98.

Shell cylindrical-ovate, rather obtuse, shining, pellucid; whorls 6–7; aperture semi-oval, mostly narrowed, by folds; peristome thickened, and reflexed.

Animal, with four distinct tentacles.

L. pentodon, Say.

Vertigo pentodon, Say, Jour. Acad. Phila. F. S., II., 1822.

L. pentodon.—[B. & B.]
Fig. 99.

Jaw of L. pentodon.—[Morse.]
Fig. 100.

Shell ovate, umbilicate, whitish; whorls 5; apex sub-acute; aperture oblique, semicircular; teeth 5, or more; lip not reflected; one prominent tooth on the pillar lip; the outer lip callous, with a series of 4 to 7 small teeth; lingual membrane, with 64 rows of 10-1-10 teeth,

Lingual Dentition of L. pentodon.—[Morse.]
Fig. 101.

each; buccal plate slightly arcuate, wrinkled longitu-dinally; with the cutting edge minutely notched. H. 2, nearly, W. 1, mill.

Station, among grass, and under boards and rubbish. Chester County; frequent.

L. marginata, Say.

Cyclostoma marginata, Say, Jour. Acad. Phila. F. S., II., 1821.

L. marginata.
[B. & B.]
Fig. 102.

Shell small cylindric-ovate, acute, pale horn color; whorls 6, convex, the last not much enlarged, and less than half the length of the shell; aperture rounded, with the margin reflected, flattened, white. H. 4½, W. 2, mill.

Station, in grass fields, and often under stones, and loose bark. Chester County; common.

L. fallax, SAY.

L. fallax.
Fig. 103.

Pupa fallax, Say, Jour. Acad. Phila. F. S., V., 1825.

Shell conical, acute; suture deep; whorls 6, the last, one-third wider than the preceding one, and more than half the entire length; aperture oval; the margin sub-revolute, but not flattened, unicolored; columella rectilinear, longitudinal. H. 4½, W. 2½. mill.

Station, under bark, near Philadelphia. (Tryon.)

L. armifera, SAY.

L. armifera, en-
larged. — [B. & B.]
Fig. 104.

Pupa armifera, Say, Jour. Acad. Phila. F. S., II., 1821.

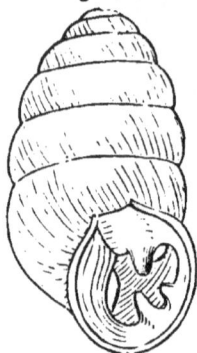

Shell oblong-oval; whorls 6; lip reflected, white, almost completing the circle; aperture nearly circular, cup-form, with four, or more, teeth — one on the pillar lip, large, prominent, one near the base, and two or three smaller ones, on the outer lip; umbilicus small. H. 3½, W. 2, mill.

Station, among grass, and under chips and stones, near streams. New Garden, Chester County; common.

L. contracta, SAY.

Pupa contracta, Say, Jour. Acad. Phila. F. S., II., 1822.

Shell whitish, cylindro-conic; whorls 5; lip thickened, subreflected, white, its extremities *L. contracta.* [B. & B.] Fig. 106. united; aperture triangular, funnel-shaped, with four teeth — one very large on the columella, a small one on the margin of the outer lip, and two larger ones, deep in the narrow throat. H. 2½, W. 1, mill.

Station, with the preceding. Chester County; abundant.

L. corticaria, SAY.

Pupa corticaria, Say, Nich. Encyc., Amer. Ed., 1817.

Shell cylindrical, obtuse, shining, whitish; *L. corticaria.* [B. & B.] Fig. 106. whorls 5; aperture lateral, sub-orbicular; lip reflected; a tooth on the pillar lip, and an internal tubercle, near the very small umbilicus. H. 2½, W. 1, mill.

Station, under the loose bark of trees, near the earth. New Garden, and near West Chester, Chester County; rare.

Genus VERTIGO, MÜLLER, 1774.

Shell rimate, ovate; apex obtuse; whorls 5–6, rounded; aperture semi-oval, with 4–7 folds; peristome not much expanded; lip white.

Animal with lappets on each side of the head; inferior tentacles wanting.

V. ovata, SAY.

Vertigo ovata, Say, Jour. Acad. Phila. F. S., II., 1822.

Shell ovate-conic, ventricose, dark-amber colored; whorls 5, the last much inflated, upper ones sub-acute; teeth six — two on the transverse lip, two on the columellar margin, and two on the outer margin ; outer lip

V. ovata. — [B. & B.]
Fig. 107.

Jaw of V. ovata. — [Morse.]
Fig. 108.

forming two segments of circles; umbilicus expanded.
H. 1¾, W. 1, mill.

Lingual Dentition of V. ovata. — [Morse.]
Fig. 109.

Station, in lawns and pastures, or among dead leaves,
chips, etc., etc. Chester County; abundant.

V. decora, GOULD.

Pupa decora, Gould, Proc. Bost. Soct. Nat. Hist., II.,
1842.

V. decora, enlarged. [Gould.] Fig. 110.

Shell minute, ovately cylindrical, amber
colored; whorls 6, well rounded, striated;
apex obtuse; suture deep, umbilicate; aperture small, rounded, with four teeth — one
on the parietal wall, one on the columella,
and two others, on the outer peritreme,
forming the arms of a cross. H. 2½, W. 1¼,
mill.

Station, among grass, in company with
V. ovata. West Chester, Chester County; rare.

V. Gouldii, BINNEY.

Pupa Gouldii, Binn., Proc. Bost. Soct. Nat. Hist., I., 1843.

V. Gouldii.
[B. & B.]
Fig. 111.

Shell very small, ovate-cylindric, chestnut-brown; whorls 4; apex obtuse; aperture lateral, with four white, prominent, teeth — one on the transverse margin, two on the umbilical margin, and two on the thick, sub-reflected lip; umbilicus closed. H. 1½, W. 1, mill.

Station, among grass, and rubbish. New Garden, Chester County; common.

V. milium, GOULD.

Pupa milium, Gould, Bost. Jour. Nat. Hist., III., 1840.

V. milium. —[B. & B.]
Fig. 112.

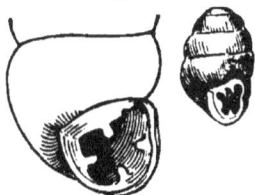

Shell very minute, chestnut col-ored, cylindric-oval; whorls 5, fine-striate; lip reflected, white; aperture lateral, half as wide as the last whorl, composed of two curves; teeth six — two on the transverse margin, two on the umbilical side, and two on the outer lip — that at the junction of the two curves being longest; umbilicus large. H. 1, W. two-thirds, mill.

Station, among grass, leaves, and under chips. New Garden, Chester County; common.

OBS. — This is one of our smallest shells. Prof. Adams says, "twelve mature specimens weighed less than one-sixteenth of a grain." 2100 shells, weighed by us, weigh only ten grains, when fully dry; which is equal to 210 shells to a grain, including the dried animals.

Sub-order LIMNOPHILA.

INOPERCULATA.

Operculum wanting.

Animal mostly fluviatile, sometimes amphibious; tentacles two, flattened, or subcylindrical, simply contractile; eyes sessile, at the base of the tentacles.

OBS. — Some of the animals of this order are considered *terrestrial*, as they are only found in wet places, within reach of the tides. They serve as the connecting link between the *Geophila* and *Limnophila*.

Family AURICULIDÆ.

Shell spiral, with a horny epidermis; aperture elongated; inner lip strongly folded; outer, frequently dentate.

Animal with subcylindrical, contractile, tentacles; with the eyes at their inner base; usually frequenting salt marshes.

Genus CARYCHIUM, MÜLLER, 1774.

Animal of Carychium exiguium.
Fig. 113.

Shell elongated, very thin, transparent; aperture with one columellar tooth; peristome expanded, its extremities not approximate; the right one, with a small columellar fold.

C. exiguium, SAY.
Pupa exigua, Say, Jour. Acad. Phila. F. S., II., 1822.

Shell very small, cylindrical, tapering at both ends, white, shining; whorls 5–6, very oblique; aperture oblique, oval, with a plait, on the columella, and a slight one near to the open umbilicus; lip thick, reflected, and flattened; lingual membrane with the rows of teeth slightly bent; buccal plate plain, slightly arched. H. 1½, W. ¾, mill.

C. exiguium.
Fig. 114.

Lingual Dentition of C. exiguium. — [Morse.]
Fig. 115.

Station, among chips, moss, etc., almost in the water. New Garden, Chester County; frequent.

Family LIMNAEIDÆ.

Shell variable, thin, horn color, usually with a slight fold on the columella; lip simple, acute.

Animal, muzzle short, broad, dilated; tentacles contractile, flattened, subulate; eyes at their inner base; respiratory orifice on the right side.

Station, fresh water; rising to the surface to breathe.

OBS. — The *Limnacidæ* are generally sluggish animals; often preferring stagnant pools to clear running streams. They are herbivorous, feeding on the small confervoid plants which everywhere abound, in places which they inhabit.

Their breeding season commences in the spring, and extends to midsummer. During this period, they are more readily found, and captured. In early autumn, they fill the pulmonary cavity with air, and soon disappear beneath the mud; still penetrating deeper, as the cold increases.

6

DIAGRAM OF THE SUB-FAMILIES, GENERA, SUB-GENERA, AND
SPECIES, OF THE FAMILY LIMNAEIDÆ.

FAMILY.	SUB-FAMILY.	GENERA.	SPECIES.
LIMNAEIDÆ.	LIMNAEINÆ.	RADIX.	columella, macrostoma.
		LIMNOPHYSA.	catascopium, ·elodes, reflexa, desidiosa, humilis, caperata.
	PHYSINÆ.	PHYSA.	ancillaria, heterostropha, gyrina, lata.
		BULINUS.	hypnorum.
	PLANORBINÆ.	SUB-GENUS. PLANORBELLA.	campanulatus.
		SUB-GENUS. HELISOMA.	trivolvis, var. fallax, bicarinatus.
		SUB-GENUS. MENETUS. .	exacutus.
		SUB-GENUS. GYRAULIS.	deflectus, parvus, albus, dilatatus.
		SUB-GENUS. PLANORBULA.	armigera.
	ANCYLINÆ.	GENUS. ANCYLUS.	rivularis, tardus.

Genus LIMNÆA, LAMARCK, 1799.

Animal of Limnæa desidiosa. — [W. G. B.]

Fig. 116.

Shell, dextral, spiral, oblong, or ovate, translucent,
horn color; spire acute; last whorl ventricose; aperture
large, wide, rounded below; inner lip with an oblique
fold; outer lip simple.

Animal, tentacles flattened, triangular; mantle thickened in front; foot short, rounded.

Sub-genus RADIX, MONTFORT, 1810.

Shell, sub-ovate, ventricose; aperture more than half
the length of the shell, greatly expanded.

R. columella, SAY.

Lymnæa columella, Say, Jour. Acad. Phila. F. S., I., 1817.

R. columella. — [W. G. B.]
Fig. 117.

Jaw of R. columella. — [W. G. B.]
Fig. 118.

Shell, thin, fragile, pale; whorls 4; spire acute; aperture expanded, ovate; columella narrowed near the base, so as to exhibit the interior of the spire. H. 15, W. 8, mill.

Lingual Dentition of R. columella. — [W. G. B.]
Fig. 119.

Station, stagnant streams and muddy pools. Chester County; common.

R. macrostoma, SAY.

Lymneus macrostomus, Say, Jour. Acad. Phila. F. S., II., 1821.

Shell, sub-oval, thin, and fragile, straw-yellow; whorls 5, with the lines of growth crossed by minute spiral striæ; suture impressed; spire acute, one-third the length of the aperture, which is large and expanded; outer peritreme slightly reflexed and forming (with a thin coat of calcareous matter on the columella) a slight umbilicus. H. 20, W. 12, mill.

R. macrostoma.
[W. G. B.]
Fig. 120.

Station, ponds. Pickering Creek; Chester County.

Sub-genus **LIMNOPHYSA**, Fitzinger, 1833.

Shell, ovate-oblong; whorls rounded; spire conic, about as long as the aperture; outer lip not spreading.

L. catascopium, Say.

L. catascopium.
[W. G. B.]
Fig. 121.

Lymnæa catascopium, Say, Nich. Ency., Amer. Ed., 1817.

Shell, horn color, or reddish; whorls 4–5, first large; the spire darker, acute; aperture large, oval, less than three-fourths the length of the shell. H. 18, W. 12, mill.

Lingual Dentition of L. catascopium. — [W. G. B.]
Fig. 122.

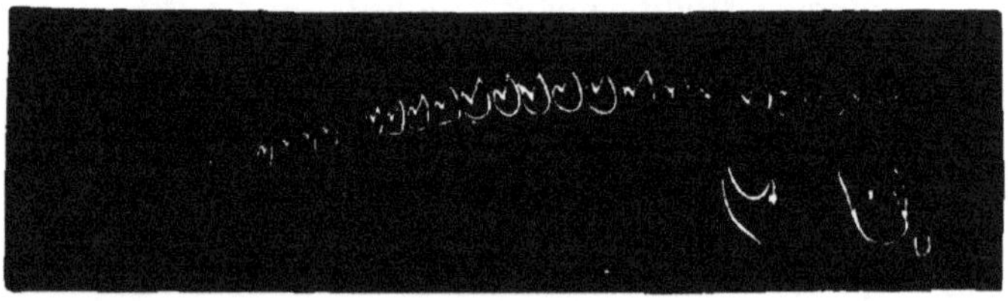

Station, Delaware and Schuylkill Rivers; abundant.

L. elodes, Say.

Lymneus elodes, Say, Jour. Acad. Phila. F. S., II., 1821.

L. elodes.
[W. G. B.]
Fig. 123.

Shell, oblong-conic, thin; whorls 5, regularly rounded; in some instances, the lines of growth, with the revolving striæ, produce, by their decussation, numerous small facets; suture moderate; umbilicus slight; aperture sub-oval, less than half the length, often reddish, and slightly thickened, within. H. 20, W. 9, mill.

Station, Schuylkill River; Chester County.

Obs. — The variable character of this species is doubtless, in a measure, due to the differences of station.

Whether (as has been alleged) L. elodes, Say, and L. catascopium, Say, are mutually convertible species, we are, at present, unable to decide.

L. reflexa, SAY.

Lymneus reflexus, Say, Jour. Acad. Phila. F.S., II., 1821.

Shell, much elongated, pale brownish; whorls 6, oblique; spire one and a half the length of the aperture, slightly reflected from the middle; two or three terminal whorls vitreous; aperture narrow; lip with a pale margin, and colored sub-margin. H. 30, W. 10, mill.

L. reflexa. [W. G. B.] Fig. 124.

Station, in lakes and ponds. New Garden, Chester County.

L. desidiosa, SAY.

Lymneus desidiosus, Say, Jour. Acad. Phila. F. S., II., 1821.

Shell, oblong, sub-conic, thin, brownish, or light ochraceous; whorls 5, convex; suture deep; lines of accretion coarse, with a tendency to form facets, on the body whorl; spire rapidly attenuated to an acute point, as long as the aperture; columellar fold slight; umbilicus small. H. 8, W. 4½, mill.

L. desidiosa. [W. G. B.] Fig. 125.

Station, ponds and streams. Chester County; common.

L. humilis, SAY.

Lymneus humilis, Say, Jour. Acad. Phila. F.S., II., 1822.

Shell, ovate-conic, thin, translucent; whorls 5–6; aperture equal to the spire, with a calcareous deposit on the pillar lip; umbilicus distinct; color yellowish, or reddish-white. H. 6, W. 3, mill.

L. humilis. [W. G. B.] Fig. 126.

Station, sluggish rivulets. Chester County; common.

L. caperata, SAY.

L. caperata.
[W. G. B.]
Fig. 127.

Lymneus caperatus, Say, New Harm. Diss., II., 1829.

Shell, elongate-oval, yellowish horn color; spire acute, equalling the length of the aperture; whorls with five equilateral revolving lines; suture impressed; aperture rather dilated. H. 26, W. 10, mill.

Station, Schuylkill River; common.

Genus PHYSA, DRAPARNAUD, 1801.

Animal of Physa.—[W. G. B.]
Fig. 128.

Shell, sinistral, oblong, thin, polished; spire acute; aperture oval, rounded below, not dilated; inner lip formed by a deposit on the last whorl; outer one acute.

Animal, tentacles slender, setaceous; mantle digitate; foot long, posteriorly, acuminate.

P. ancillaria, SAY, Jour. Acad. Phila. F. S., V., 1825.

P. ancillaria.
[W. G. B.]
Fig. 129.

Shell, sub-globose, brownish - yellow; whorls 4, rapidly attenuated; spire hardly elevated above the general curve; aperture dilated, but little shorter than the shell; lip thickened on its inner margin. H. 16, W. 12, mill.

Station, Brandywine Creek, Chester County.

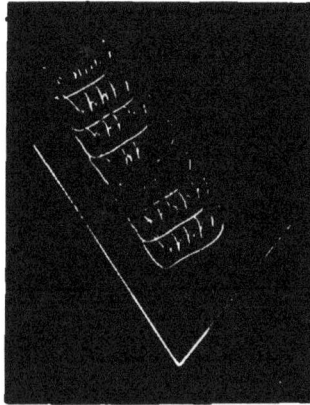

P. heterostropha, SAY, Jour. Acad. Phila. F. S., II., 1821.

Shell sub-ovate, pale yellow to chestnut; whorls 4, the first large, the rest very small; apex abrupt, acute; aperture oval, three-fourths the length of the shell; lip slightly thickened, within, pearly. H. 16, W. 9, mill.

P. heterostropha.
[W. G. B.]
Fig. 131.

Station, springs and small streams. Chester County; very abundant.

P. gyrina, SAY, Jour. Acad. Phila. F. S., II., 1821.

Shell, oblong; whorls 5–6, gradually acuminated; aperture a little more than half the length of the shell; outer lip thickened within the margin. H. 17, W. 10, mill.

P. gyrina.
[W. G. B.]
Fig. 132.

Station, streams. Western Pennsylvania.

OBS. — Sometimes confounded with the preceding species.

P. lata, Tryon, Contin. F. W. Moll. U. S.

P. lata.
[Tryon.]
Fig. 133.

Shell, very fragile, light horn color, waxy, irregularly striate; spire moderately elevated; whorls convex; apex acute; suture well impressed; body inflated; aperture rather large; columellar lip turned to the right, very narrow, distinctly folded. H. 10.5, W. 7, mill.

Station, Juniata River, Hollidaysburg, Pennsylvania.

Obs. — This species not having fallen under our observation, the figure and description have been taken from the recent admirable continuation of Haldeman's Monograph of the Fresh Water Univalve Mollusca, by G. W. Tryon, Jr., who observes that "this species has very much the form of *P. heterostropha*, Say, but is rather more ventricose, much thinner, and the surface exhibits a peculiar, glimmering lustre."

Genus BULINUS, Adanson, 1757.

Animal of Bulinus.
Fig. 134.

Shell, sinistral, elongated, polished, thin, acuminate; aperture narrow, produced anteriorly.

Animal, tentacles filiform, setaceous; mantle simple, not fringed, and not reflected over the shell, as in Physa.

B. hypnorum, Linn.
Physa hypnorum, Linn., Syst. Nat., 1758.

B. hypnorum.
[W. G. B.]
Fig. 135.

Shell, oblong, fragile, diaphanous; whorls 6; spire acute; aperture not dilated, attenuated above, half the length of the shell; columella narrowed at the base. H. 17, W. 7, mill.

Station, in flowing streams, and stagnant pools. New Garden, Chester County; rare.

Genus PLANORBIS, Guettard, 1756.

Animal of P. bicarinatus. — [W. G. B.]
Fig. 136.

Shell, dextral, discoidal; spire depressed; whorls numerous, visible on both sides; aperture crescentic, or transverse-oval; peristome thin, the upper margin produced.

Animal, tentacles slender, filiform; foot short, ovate.

Sub-genus PLANORBELLA, Haldeman, 1842.

Shell, with few whorls; aperture campanulate, or expanded, prominent.

P. campanulatus, Say.

Planorbis campanulatis, Say, Jour. Acad. Phila. F. S., II., 1821.

Shell, with 4 whorls, longer than wide; spire plane, or nearly so; body whorl abruptly dilated, near the aperture, sub-campanulate; umbilicus profound. H. 7, W. 17–12, mill., in its greater and less diameters.

P. campanulatus.
[W. G. B.]
Fig. 137.

Station, in rivers and ponds. Delaware Water Gap, Pennsylvania.

Sub-genus HELISOMA, Swainson, 1840.

Shell, ventricose; spire sunk below the body whorl; whorls few, often angulated.

H. trivolvis, SAY.

H. trivolvis.
[W. G. B.]
Fig. 138.

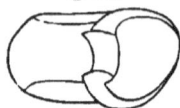

Planorbis trivolvis, Say, Nich. Encyc.,
Amer. Ed., 1816.

Shell, pale brownish-yellow, sub-carinate above and beneath, especially the
inner whorls; whorls 3–4, fine ridge-
striate; spire concave; aperture large;
lip thickened, reddish within, vaulted on
its shorter side; umbilicus wide, shallow.
H. 8, W. 17–13, mill.

Station, ponds and streams. Chester
County; common.

H. trivolvis, Var. fallax, HALD., Lim., 1842.

H. trivolvis, Var. fallax. — [W. G. B.]
Fig. 139.

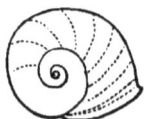

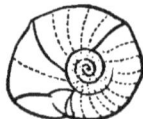

Shell, thin, translucent, carinate below; umbilicus
very shallow; whorls 3, obsolete-striate; lip acute,
margined; aperture ovate. H. 7, W. 13–10, mill.

Station, Schuylkill River, Chester County.

H. bicarinata, SAY.

H. bicarinata.
[W. G. B.]
Fig. 140.

Planorbis bicarinatus, Say, Nich. Encyc.,
Amer. Ed., 1816.

Shell, pale brownish-yellow; whorls 3,
carinate on both sides; aperture large; lip
sub-revolute, arched above and below;
within chestnut, with two arched lines an-
swering to the carina. H. 6, W. 12–9,
mill.

Station, ponds and streams. Chester County; common.

Sub-genus MENETUS, H. and A. ADAMS, 1853.

Shell, depressed; whorls rapidly increasing; periphery angulated.

M. exacutus, SAY.

Planorbis exacutus, Say, Jour. Acad. Phila. F. S., II., 1821.

M. exacutus. —[W. G. B.]
Fig. 141.

M. exacutus (abnormal).
Fig. 142.

Shell, small, depressed; whorls 4, wider than long, carinated on the circumference; umbilicus broad, and deep; aperture below the carina, and angulated by it. H. 1½, W. 4, mill.

Station, ponds. Pickering Creek, Chester County.

Sub-genus GYRAULUS, AGASSIZ, 1837.

Shell, orbicular above, flat beneath; whorls few, rapidly increasing.

G. deflectus, SAY.

Planorbis deflectus, Say, Long's Exped., II., 1824.

G. deflectus.
[W. G. B.]
Fig. 143.

Shell, small, depressed; whorls nearly 5, flattened, smooth, fine striate; margin obtusely carinate; last whorl deflected downwards near the aperture; which is large and very oblique. H. 2, W. 6, mill.

Station, ponds and rivulets. Pickering Creek, Chester County; frequent.

G. parvus, SAY.

Planorbis parvus, Say, Nich. Encyc., Amer. Ed., 1816.

G. parvus. Shell, very small, horn color; whorls 4,
[W. G. B.] fine striate; flat above, with the centre im-
Fig. 144. pressed; beneath, concave; periphery often
sub-carinate; aperture oval, oblique, longer
than wide. H. ¾, W. 2½, mill.

Station, with the preceding species. Chester County;
common.

G. albus, MÜLLER.

Planorbis albus, Müll., Hist. Vermet., II., 1773.

G. albus. Shell, small, pale yellowish, or white; whorls
[W. G. B.] 3–4, the outer rapidly increasing, and marked
Fig. 145. by revolving lines, with stiff hair; both sides
concave, the lower most so; aperture sub-oval,
very oblique. H. 1¼, W. 5, mill.

Station, tide marsh ditches. Delaware and
Schuylkill Rivers, Philadelphia.

G. dilatatus, GOULD.

Planorbis dilatatus, Gould, Invert. Mass., 1841.

G. dilatatus. Shell, very small, yellowish-green; whorls
[W. G. B.] 3, outer one carinate; slightly convex above;
Fig. 146. closely umbilicate, beneath; aperture large,
oblique, trumpet shaped. H. 1, W. 3, mill.

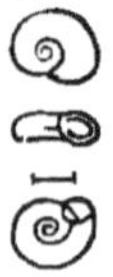

Stations, ponds. Pickering Creek, Chester
County; common.

Sub-genus PLANORBULA, HALDEMAN, 1842.

Shell, with the aperture furnished with dentiform
plicæ, not forming open partitions.

P. armigera, SAY.

Planorbis armigerus, Say, Jour. Acad. Phila. F. S., II., 1821.

Shell, light brown, polished; whorls 4, sub-cylindrical, and sub-carinate below; umbilicus wide, and deep; aperture rounded, with six white teeth, far within the throat; the largest thin, oblique, running back from the left to the right side; on the left of this is a small one, and around the vault four others; the teeth sometimes wanting. H. 2½, W. 6½, mill.

P. armigera.
[W. G. B.]
Fig. 147.

Aperture of P. armigera.
Fig. 148.

Station, Schuylkill River, and tributaries. Chester County.

Genus ANCYLUS, GEOFFROY, 1767.

Shell, thin, patelliform, non-spiral; apex directed to the right; aperture wide; peritreme continuous, simple, entire.

Animal of Ancylus.
[W. G. B.]
Fig. 149.

Animal, tentacles triangular; mantle included; pulmonary orifice protected by a branchial appendage; foot large.

A. rivularis, SAY, Jour. Acad. Phila. F. S., II., 1817.

Shell, delicate, depressed-conic; apex obtuse nearer to, and leaning towards, one side, and one end; aperture oval, narrower at one end, entire; within milk white. H. 1½, W. 3, mill.

A. rivularis.
[W. G. B.]
Fig. 150.

Station, on dead shells, and stones, in streams. Chester County; frequent.

A. tardus, SAY, New Harm. Diss., V., 1830.

A. tardus.
[W. G. B.]
Fig. 151.

Shell, conic, white, opaque; apex almost terminal, inclining backwards; anterior slope rounded; posterior straight, almost perpendicular; sides equal, steep, curved; aperture broad, oval, not narrowed at one end. H. 3, L. 5, W. 3½, mill.

Station, in stagnant pools. New Garden, Chester County; rare.

OPERCULATA.

With an opercle.

Order PECTINIBRANCHIATA.

Animal, with the gills arranged in numerous, parallel, lamina.

Sub-order ROSTRIFERA.

Animal, fluviatile or marine; head produced; rostrum contractile; tentacles subulate; operculate.

Family VIVIPARIDÆ.

Shell, conoid; whorls convex; aperture ovate or subrotund; epidermis green, or olivaceous.

Animal, viviparous; rostrum small, simple, truncated, extending slightly beyond the shell; tentacles short, subulate, the right one, on the male, as large as the rostrum; eyes on peduncles at the exterior base of the tentacles; foot large; operculum corneous, concentric, sometimes with a spiral nucleus.

DIAGRAM OF THE GENERA, AND SPECIES, OF THE FAMILIES VIVI-
PARIDÆ, VALVATIDÆ, AMNICOLIDÆ, AND STREPOMATIDÆ.

FAMILY.	GENERA.	SPECIES.
VIVIPARIDÆ.	{ MELANTHO.	decisa.
	{ LIOPLAX.	sub-carinata.
VALVATIDÆ.	VALVATA.	tricarinata.
AMNICOLIDÆ.	AMNICOLA.	{ limosa, decisa, grana.
	BYTHINELLA.	Nickliniana.
	POMATIOPSIS.	{ lapidaria, lustrica.
	SOMATOGYRUS.	altilis.
STREPOMATIDÆ.	GONIOBASIS.	{ Virginica, var. multilineata.
	ANCULOSA.	dissimilis.

Sub-genus MELANTHO, BOWDITCH, 1822.

Animal of Melantho. —[W. G. B.]
Fig. 152.

Shell, thick, solid, ovate imperforate; whorls rounded, smooth; peristome simple, continuous.

Animal, with the foot broad, and thin, produced beyond the rostrum, which is short and truncated; operculum concentric.

M. decisa, SAY.

Paludina decisa, Say, Nich. Encyc., 1st Amer. Edit., 1817.

Shell, elongate ovate; whorls 4, wrinkled across, and with minute revolving striæ; aperture ovate, more than half the length of the last whorl; epidermis green;

within bluish-white; operculum with the centre depressed.   H. 28, W. 15, aperture H. 15, W. 9, mill.

M. decisa. — [W. G. B.]
Fig. 153.

Station, creeks and ponds.   Chester County; abundant.

Lingual Dentition of M. decisa.
Fig. 154.

Operculum of M. decisa.
Fig. 155.

Genus LIOPLAX, Troschel, 1857.

Animal of L. subcarinata, male and female. — [W. G. B.]
Fig. 156.

Female.   Male.

Shell, thin, ovate, turreted; whorls rounded, carinate.

Animal, foot large, greatly produced beyond the rostrum, narrow, and rounded behind, truncated in front; rostrum short; tentacles broader, and shorter, than in Melantho; the right one, in the male, only one-third the length of the left; operculum with a sub-spiral nucleus.

L. subcarinata, SAY.

Paludina subcarinata, Say, Nich. Encyc., 3d Amer. Edit., 1819.

Shell, whorls 3 or 4, reticulated, with striæ, and wrinkles; apex truncate, and re-entering; suture deeply impressed; aperture oval, more than half the length of the shell; carina, two, or three, or absent; color brownish - green; within bluish-white. H. 15, W. 9, mill.

L. subcarinata.
[W. G. B.]
Fig. 157.

Station, Schuylkill River, Chester County; abundant.

Lingual Dentition of L. sub-carinata. — [W. G. B.]
Fig. 158.

Operculum of L. sub-carinata.—[W. G. B.]
Fig. 159.

OBS. — Our shell differs, somewhat, from the same species found in Western waters.

Family VALVATIDÆ.

Animal of V. tricarinata. — [W. G. B.]
Fig. 160.

Shell, spiral, turbinate, or discoid; peristome entire.

7 *

Animal, rostrum produced; tentacles cylindrical; eyes sessile, at their extreme bases; mantle simple in front; gill exposed, plumose, lamina pinnate, twisted, protected by a respiratory lobe; foot bilobed in front; operculum corneous, spiral.

Genus VALVATA, O. F. MÜLLER, 1774.

Shell, sub - discoid, or conoid, umbilicated, thin; whorls cylindrical, or keeled; peristome circular, continuous.

V. tricarinata, SAY.

Cyclostoma tricarinata, Say, Jour. Acad. Phila., I., 1817.

V. tricarinata,
[W. G. B.]
Fig. 161.

Shell, with three whorls, and three prominent carina, or revolving lines, one of which is on the upper edge, one on the lower edge of the whorl, and one beneath the base; suture canaliculate; umbilicus large. W. 5, mill.

Station, Schuylkill River, and tributaries. Chester County; abundant.

Operculum of V. tricarinata.
[W. G. B.]
Fig. 162.

Lingual Dentition of V. tricarinata.
[W. G. B.]
Fig. 163.

OBS. — The ova are deposited from May to midsummer, in globose, gelatinous masses, of 10 to 30, and of a green color; these are hatched in fourteen or fifteen days; like Limnea, and Physa, they possess the power of swimming in an inverted position, along the surface of water.

Family AMNICOLIDÆ.

Shell, turbinate, or elongate-turreted, smooth, perforate, or umbilicate; peristome continuous; shell often covered with blackish incrustations.

Animal, head proboscidiform, extending beyond the foot; tentacles cylindrical, elongated; eyes sessile at their outer bases; foot oval, truncated in front, and rounded, or pointed, behind; verge exserted, behind the right tentacle; operculum corneous, sub-spiral.

Genus AMNICOLA, GOULD, and HALDEMAN, 1840.

Animal of A. limosa, enlarged. — [Stimpson.]
Fig. 164.

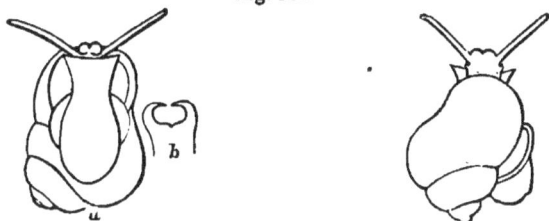

Shell, small, sub-ovate, thin, smooth, perforate; spire short; aperture broad-ovate; peristome continuous.

Animal, foot short, broad, expanded, and rounded, behind, auricled in front; rostrum short; tentacles cylindrical, blunt; verge bifid, with a globular base; opercle thin, sub-spiral.

A. limosa, SAY.

Paludina limosa, Say, Jour. Acad. Phila., I., 1817.

A. limosa.
[W. G. B.]
Fig. 165.

Shell, conic, sub - umbilicate, dark horn color; epidermis wrinkled; aperture ovate-orbicular; suture impressed. H. 3, W. 2, mill.

Station, Schuylkill River, and Brandywine Creek, Chester County; common.

Lingual Dentition of A. limosa. — [Stimp.]
Fig. 166.

Operculum of A. limosa. — [Stimp.]
Fig. 167.

A. decisa, HALDEMAN, Mon. Lim., 1842.

A. decisa.
[W. G. B.]
Fig. 168.

Shell, short, conical, smooth, shining; whorls 5; base slightly perforated; aperture dilated, semicircular; color pale green; slightly translucent. H. 4, W. 2½, mill.

Station, Brandywine Creek, Chester County; not common.

A. grana, SAY.

Paludina grana, Say, Jour. Acad. Phila., II., 1822.

A. grana.
[W. G. B.]
Fig. 169.

Shell, very small, conic-ovate; whorls 5, convex, smooth; aperture orbicular, sub-angular; umbilicate. L. 2, W. 1½, mill.

Station, in ponds, on fallen leaves. Pickering Creek, Chester County; abundant.

Genus BYTHINELLA, MOQUIN TANDON, 1851.

Shell, elongated-ovate, turreted; umbilicus generally closed; apex obtuse; aperture oval, or rounded, slightly thickened.

Animal, with the foot rather long, narrow, and rounded behind; tentacles setaceous and pointed; verge bifid; the longer branch not coiled about the shorter one; operculum sub-spiral; concealing the posterior part of the foot.

B. Nickliniana, LEA.

Paludina Nickliniana, Lea, Tr. Amer. Phil. Soc., VI., 1839.

Shell, green, with four convex, smooth, whorls, turrited; apex obtuse; aperture ovate. H. 3½, W. 1½, mill.

B.Nickliniana.
[W. G. B.]
Fig. 170.

Station, tributaries of the Susquehanna River, Chester County; common.

Lingual Dentition of B. Nickliniana. — [Stimp.]
Fig. 171.

OBS. — *B. attenuata,* Haldeman, is the more perfect specimen of this species, occurring in spring heads, and still water.

Genus POMATIOPSIS, TRYON, 1862.

Animal of P. lapidaria. — [Stimp.]
Fig. 172.

Shell, small, thin, smooth, elongated, turreted, sub-umbilicate; aperture ovate; peristome reflected.

Animal, tentacles short, subulate, pointed; rostrum large, longer than the tentacles, and transversely wrinkled; foot broad, with lateral sinuses; verge very large, flattened, broad, spiral; gills pectinate; opercle thin, sub-spiral, amphibious; progression stepping.

F

P. lapidaria, Say.

Cyclostoma lapidaria, Say, Jour. Acad. Phila., I., 1817.

P. lapidaria.
[W. G. B.]
Fig. 173.

Shell, sub-umbilicate; whorls 6, obsolete-ly, and transversely, wrinkled; aperture longitudinally ovate-orbicular, more than one-third the length of the shell. H. 6, W. 2½, mill.

Lingual Dentition of P. lapidaria. — [Stimp.]
Fig. 174.

Station, under stones, near water. New Garden, Chester County; not common.

P. lustrica, Say.

Paludina lustrica, Say, Jour. Phila. Acad., II., 1821.

P. lustrica.
[W. G. B.]
Fig. 175.

Shell, conic; whorls convex; aperture oval-orbicular; upper margin of the lip not appressed to the last whorl; umbilicus rather large, rounded. H. 5, W. ¾, mill.

Station, shores of the Delaware River. (Tryon.)

Obs. — Possibly, the young of the preceding species.

Genus SOMATOGYRUS, Gill, 1863.

Shell, short, sub-conic, striate, thin, or moderately thickened; spire small; aperture large, broad-ovate, oblique; operculum thin, sub-spiral.

Animal, with the foot oblong, broadly rounded behind, and auricled in front; rostrum broad; tentacles long, slender, and pointed; eyes at the outer sides of tubercles, at the outer bases of the tentacles; verge small, simple, lunate.

S. altilis, LEA.

Melania altilis, Lea, Proc. Amer. Phila. Soct., II., 1841.

S. altilis.
[W. G. B.]
Fig. 176.

Shell, smooth, sub-globose, thick, pale horn color; spire short; whorls 4, obtusely angular above; aperture large, nearly round, white. H. 8, W. 6, mill.

Lingual Dentition of S. altilis, with the rachidian tooth greatly enlarged. — [Stimp.]
Fig. 177.

Station, Delaware and Susquehanna Rivers; abundant.

Family STREPOMATIDÆ.

Shell, elongate-conical, or turbinate; peristome not entire.

Animal, rostrum long; foot oval; tentacles subulate; eyes at their external bases; margin of the mantle not digitate; opercle corneous, semi-concentric, or pauci-spiral; oviparous.

Genus GONIOBASIS, LEA, 1862.

Shell, thick, elongate-conical; aperture sub-rhomboid, higher than wide; columella not twisted; base sub-angular, without a channel.

G. Virginica, GMELIN.

G. Virginica.
Fig. 178.
Buccinum Virginicum, Gmelin, Syst. Nat., I., 1788.

Shell, rather thin, smooth, elongated; whorls 6, or 7, slightly convex, upper ones carinate; aperture oblong-elliptical; color brown, or olivaceous, usually with two revolving, reddish bands; within bluish-white. H. 25–32, W. 10, mill.

Station, Schuylkill River, Chester County; abundant.

G. Virginica, var. multilineata, SAY, Jour. Acad. Phila., II., 1822.

Shell, resembling the preceding, but with numerous, filiform, elevated, sub-equal, lines; from ten to twenty on the body whorl.

Station, found with the preceding.

Genus ANCULOSA, SAY, 1821.

Shell, solid, ovate, or globose; spire very short; columella callous above; peristome almost as large as the whole shell, not continuous; opercle corneous, semi-concentric, or pauci-spiral.

A. dissimilis, SAY.

Paludina dissimilis, Say, Nich. Encyc., 3d Edit., 1819.

A. dissimilis.
Fig. 179.
Shell, thin, ovate, conical; whorls 3–4; body whorl large, more or less carinate in the middle, carina sometimes double, or wanting; aperture ovate; columella flattened at base, and occasionally toothed; color olive-green, or yellow. H. 12, W. 8, mill.

Station, Susquehanna River and tributaries; abundant.

Lingual dentition of A. dissimilis. — [Troschel.]
Fig. 180,

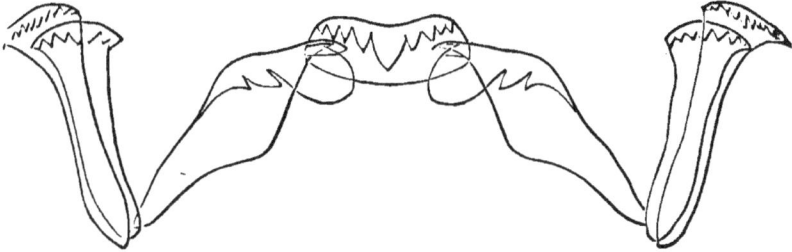

Class ACEPHALA.

In this class the shell is bivalve, the valves being united at the back, by a hinge, as in the common creek mussel.

Animal, acephalous; the mouth is a simple aperture, without teeth, or jaws, and often takes the form of a tube, reaching even beyond the shell, and accompanied by another — the vent; which are simple, or fringed.

Respiration is performed by two, or four, broad leaves, or gills, which hang down on the sides of the body, within the mantle, and are embraced by the shell, like the leaves of a book by the cover. The animal is either free, as the mussel, or attached, as the oyster.

Their food is only such animal, and vegetable, matters, as the water may chance to carry into their suctorial mouths.

They are bisexual, as the mussel; or hermaphrodite, and capable of self-impregnation, as the oyster.

Order BRANCHIFERA.

Characters the same as those of the class.

8

Family UNIONIDÆ.

Shell, transverse, inequilateral, covered with an epidermis; hinge with a simple, or divided, cardinal tooth in front, and a lamellar one running back beneath the ligament; sometimes the lamellar tooth is wanting; and sometimes both are absent.

Animal, mantle open beneath, its edges thickened, often fringed; tubes two, the vent tube above; foot thick, compressed, muscular, coriaceous.

DIAGRAM OF THE GENERA, AND SPECIES, OF THE FAMILY UNIONIDÆ.

FAMILY.	GENUS.	SPECIES.
UNIONIDÆ.	UNIO.	complanatus, radiatus, cariosus, ochraceus, Fisherianus, nasutus, heterodon.
	MARGARITANA.	margaritifera, undulata, marginata.
	ANODONTA.	fluviatilis, implicata, edentula, Tryonii.

Genus UNIO, RETZIUS, 1788.

Shell, transverse, with three deeply-impressed cicatrices; thick, or thin and fragile; teeth varying, or absent.

Animal, the mouth lips wider than long, united for two-thirds on the upper margin; mantle open; branchiæ four, the outer free posteriorly, and lies in a fold of the mantle; the inner are united to the foot anteriorly, the remainder free; foot tongue - shaped, and produced anteriorly.

Unio complanatus, SOLANDER.

Mya complanata, Solander, MSS. Portland Cat. in British Museum, Unio purpureus, Say, Barnes, and others.

Shell, heavy, ovate, or rhomboid, inequilateral, round before, sub-angular behind; base curved, hinge margin

Unio complanatus.
Fig. 181.

elevated, sloped posteriorly; beaks not prominent; epidermis dark olive-green; rayed when young; nacre varying from white to dark purple, often iridescent; cardinal teeth thick. H. 2.5, W. 4–4½, T. 1–1½, inches. Very variable.

Station, ponds and running streams. Chester County; everywhere common.

U. radiatus, LAMARCK, An. Sans. Ver., VI., 1819.

U. radiatus.
Fig. 182.

Shell, transversely ovate, compressed; anterior end

narrow, rounded; posterior, broad and rounded; dorsal margin elevated behind the hinge; inferior curved; beaks moderate; epidermis mostly greenish - yellow, with numerous broad, dark-green rays; interior yellowish, and iridescent; cardinal teeth bifid, crenulated, and oblique. H. 2–2½, W. 3–4, T. 1–1½, inches.

Station, Schuylkill River, Chester County.

U. cariosus, SAY, Nich. Encyc., 1st Amer. Ed., 1816.

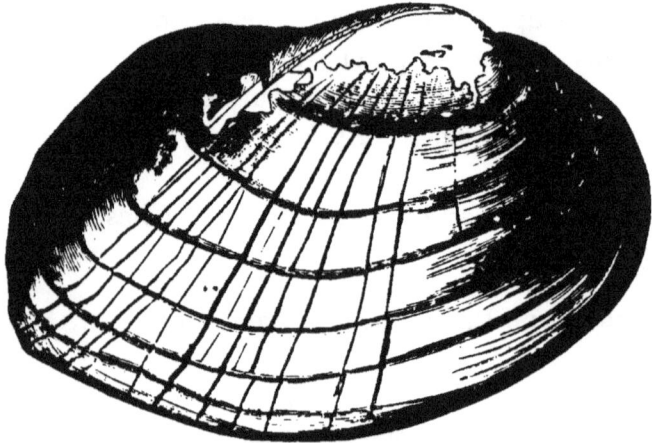

U. cariosus.
Fig. 183.

Shell, valves rather thin, sub-oval, inflated; hinge margin elevated, base curved; anterior end short, rounded, posterior narrowly rounded; the posterior dorsal slope sometimes with four, or five, fine interrupted wrinkles, and green rays; beaks at one-fourth from the anterior end; yellow or greenish; inside bluish-white; cardinal teeth long, compressed, oblique, crenate. H. 2.5, W. 3.5, L. 1.75, inches.

Station, Schuylkill, Delaware, and Susquehanna Rivers; common.

U. ochraceus, Say, Nich. Encyc., 1st Amer. Ed., 1816.

U. ochraceus.
Fig. 184.

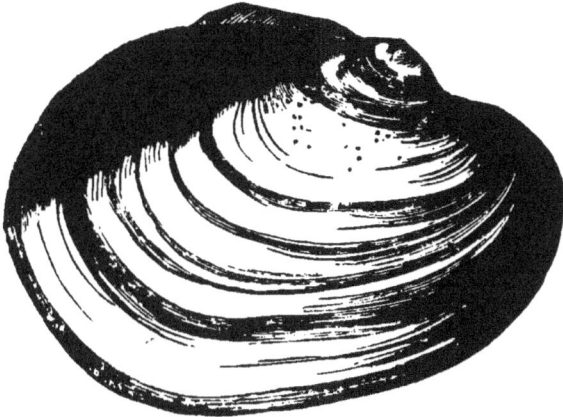

Shell, thin, thicker anteriorly, translucent, pale yellowish, or reddish; radiate with dull green, and with minute wrinkled radii, most observable on the posterior slope; sub-ovate, hinge margin rectilinear, anterior and lower, rounded, posterior oblique-truncate, angular; beaks undulated; within bluish; cardinal teeth double, much compressed, almost horizontal. H. 2, W. 3.5, T. 1.5, inches.

Station, with the preceding.

U. Tappanianus, Lea, Tr. Amer. Ph. Soct., VI., 1817.

U. Tappanianus.
Fig. 185.

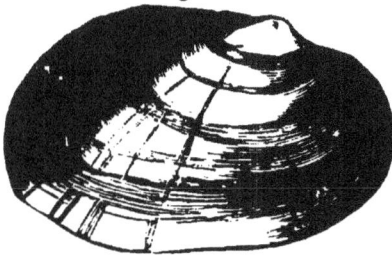

Shell, thin, sub-oval, rather inflated, elevated back of

8 *

the ligament; beaks small, undulated; yellowish-brown, with dark-green rays; nacre inclined to pale salmon color; cardinal teeth double, sometimes trifid, long, compressed, oblique. H. 1.25, W. 2.25, T. .8, inches.

Station, Schuylkill River, Chester County; common.

U. nasutus, SAY, Nich. Encyc., 1st Amer. Ed., 1816.

U. nasutus.
Fig. 186.

Shell, thin, oblong, rounded before, rostrate posteriorly; fuscous, radiate with green, behind the middle; beaks small, undulate; within bluish-white; cardinal teeth small, oblique, crenate. H. 1.25, W. 2.6, T. .75, inches.

Station, Schuylkill River, Chester County.

U. Fisherianus, LEA, Tr. Amer. Phil. Soct., VI., 1834.

U. Fisherianus.
Fig. 187.

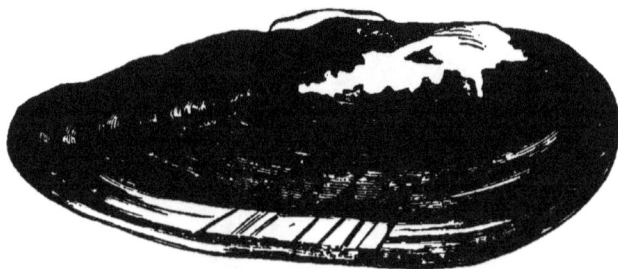

Shell, thin, compressed; beaks very near the anterior.

rounded, extremity; hinge margin elevated; base slightly curved; posteriorly sloped above, and below, to an acute angle; dark brown, or greenish; nacre hyaline-purple; cardinal teeth small, compressed, oblique. H. .85, W. 2.25, L. 4, inches.

Station, White Clay Creek, Chester County.

U. heterodon, LEA, Trans. Amer. Phil. Soct., III., 1829.

U. heterodon.
Fig. 188.

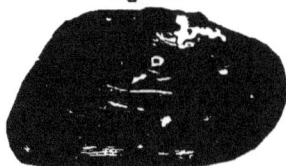

Shell, small, thin, greenish-brown, obscurely rayed; rhomboid-ovate, ventricose; rounded before; base nearly straight; dorsal margin elevated, behind the hinge; posterior margin oblique; beaks undulated, prominent; nacre bluish-white; cardinal teeth compressed, oblique; lateral tooth double in the right valve. H. .5, W. .8, T. .4, inch.

Station, Schuylkill River, Chester County; abundant.

Sub-genus MARGARITANA, SCHUMACHER, 1817.

Shell, thick, strong; cardinal teeth stout, triangular; lateral tooth absent.

Animal, gills free from the abdominal sac, and not united to the mantle, posteriorly; inferior siphon fringed, superior plane.

M. margaritifera, LINN.
Mya margaritifera, Linnaeus, Syst. Nat., 1758.
Alasmodonta, Barnes, 1823.

Shell, elongate, arcuate, cylindric-ovate ; anterior end rounded; posterior produced; ligament margin regularly

Margaritana margaritifera.
Fig. 189.

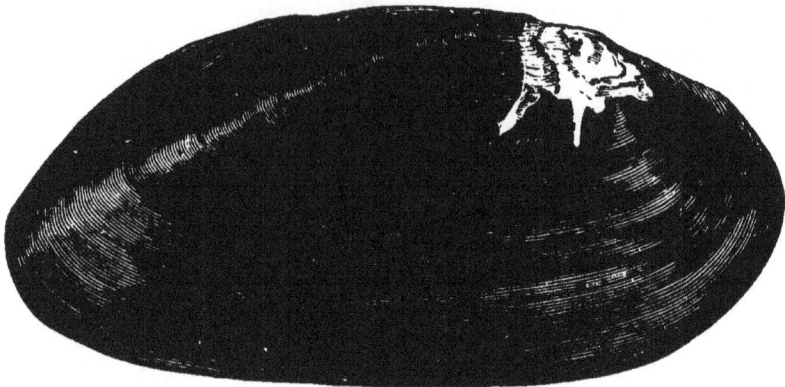

curved, and carinate behind the hinge ; base arcuated ; beaks slightly elevated; epidermis brownish-black ; nacre whitish, posteriorly iridescent ; teeth stout, grooved. H. 2.25, W. 5, T. 1.5, inches.

Station, creeks and ponds. White Clay Creek, Chester County ; not common.

M. undulata, SAY.
Unio undulata, Say, Nich. Encyc., 1st Amer. Ed., 1816.

M. undulata.
Fig. 190.

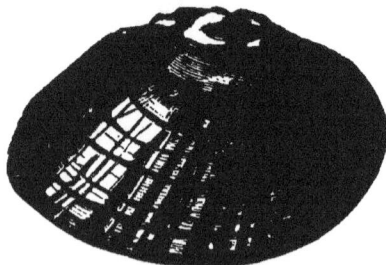

Shell, convex, elongate-oval, olivaceous, with obtuse,

concentric, wrinkles, and radiated with green; beaks prominent, undulated, usually decorticated; the anterior portion of the valves thickened internally, and whitish; posteriorly, thin, and salmon color, with iridescence; teeth stout, triangular, crenated, double in the left valve, and prolonged to the hinge margin. H. 1.6, W. 3, T. 1.25, inches.

Station, creeks and ponds. Chester County; common.

M. marginata, SAY.

Alasmodonta marginata, Say, Jour. Acad. Nat. Sci., I., 1818.

M. marginata.
Fig. 191.

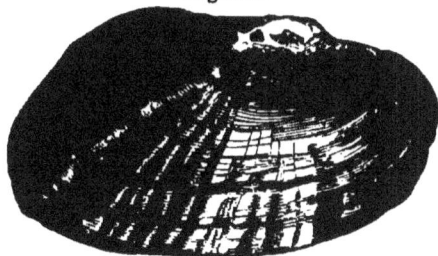

Shell, thin, transversely oblong, or sub-triangular, inflated, with the ligament slope abruptly depressed, and obliquely rugose; epidermis olive-brown, with green rays; within bluish-white; teeth compressed, oblique, nearly parallel with the anterior margin, and abrupt behind. H. 1.65, W. 2.8, T. 1.3, inches.

Station, creeks and ponds. Chester County; frequent.

Sub-genus **ANODON** (BRUGIERE), CUVIER, 1798.

Shell, equivalve, inequilateral, transverse; hinge margin linear, without teeth.

Animal, mantle margin entirely free; the outer branchia united to the mantle, as far as its extremity; the inner entirely united to the foot; foot thick, tongue-shaped, produced anteriorly.

A. fluviatilis, DILLWIN.

Mytilus fluviatilís, Dillwin, Descr. Cat. Shells, 1817.

Anodon fluviatilis.
Fig. 192.

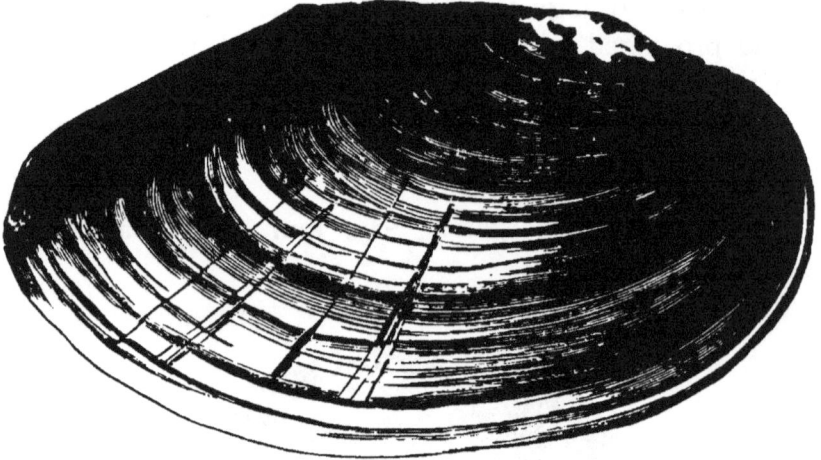

Shell, thin, fragile, oblong-ovate, ventricose; epidermis olive-brown, rayed with green; hinge margin elevated; beaks at the anterior third of the length; nacre bluish-white. H. 2.75, W. 5, T. 2, inches.

Station, in all the larger streams, especially in milldams. Chester County; abundant.

A. implicata, SAY, New Harm. Dis., II., 1829.

A. implicata.
Fig. 193.

Shell, large, thick, oblique, sub-rhomboid, ventricose; beaks prominent, undulated; epidermis dark-brown, or yellowish, without rays; surface undulated, smooth; the anterior half of the valves, near their edge, thickened internally. H. 3, W. 5.5, T. 2.3, inches.

Station, Schuylkill River, Chester County.

A. edentula, SAY.

Alasmodonta edentula, Say, New Harm. Dis., II., 1829.

A. edentula.
Fig. 194.

Shell, transverse-oval, compressed, olivaceous, rayed with green; beaks small, undulated; within bluish-white; teeth obsolete, but the peculiar curvature, beneath the beaks, shows their locality. H. 1.5, W. 2.75, T. .8, inches.

Station, running streams. Brandywine Creek, Chester County; common.

A. Tryonii, LEA, Jour. Acad. Phila. N. S., VI., 1862.

A. Tryonii.
Fig. 195.

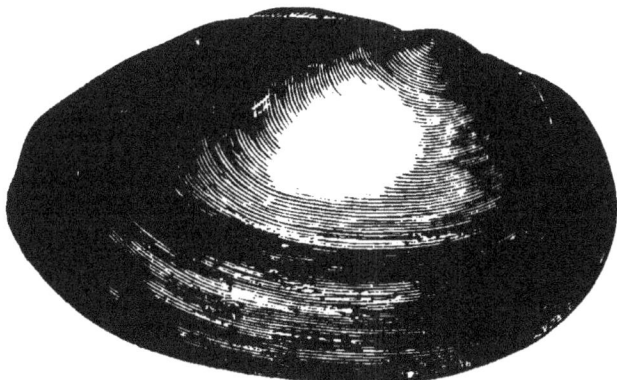

Shell, thin, smooth, elliptic-ovate, compressed, rounded anteriorly, produced and sub-angulated posteriorly; dorsal margin behind the ligament elevated, and compressed; epidermis more or less brown, or green, rays obsolete; nacre bluish-white, and iridescent. H. 1.25, W. 2.25, T. .75, inches.

Station, Schuylkill River, Chester County.

Family CORBICULIDÆ.

Shell, bivalve, oval, or sub-triangular; primary teeth two, or three, in each valve; laterals two; ligament external.

Animal, mantle lobes free in front, and at the base; siphons two; foot triangular, cylindrical, or linguiform; tentacles small, triangular, pointed; gills broad, unequal, united posteriorly.

DIAGRAM OF THE GENERA, AND SPECIES, OF THE FAMILY CORBICULIDÆ.

FAMILY.	GENUS.	SPECIES.
CORBICULIDÆ.	SPHÆRIUM.	sulcatum, striatinum, stamineum, fabalis, transversum, rosaceum, partumeium, securis.
	PISIDIUM.	virginium. variabile, abditum, compressum.

Genus SPHÆRIUM, Scopoli, 1777.

Shell, oval, nearly equilateral, beaks prominent; two primary teeth in each valve; laterals elongated; ligament on the longer slope.

Animal, with two short siphons, joined at base, oral tentacles short; foot narrow, elongated.

Obs. — The small animals included in this family

inhabit still waters, and running streams — some preferring the one, some the other. The species are widely
distributed, and their appearance so much influenced by
locality, as to produce many varieties, which have too
often been mistaken for new species, and named as such,
causing much confusion, and a needless synonymy. The
same remark will apply to many other families of fresh
water shells, especially the Strepomatidæ.

The habits of the animals of Pisidium and Sphærium
are somewhat similar, and they are often found together.
They are dioicous and viviparous. The breeding season
reaches from April to midsummer, during which period
they may be found crawling on the surface of the mud.
At a later period they bury themselves in the mud, with
the aid of the long, extensible foot, where they remain
during the ensuing winter.

S. transversum, SAY.

Cyclas transversum, Say, New Harm. Dissem., II., 1829.

Sphærium transversum. — [Prime.]
Fig. 196.

Shell, transversely oblong, inequilateral, translucent,
anterior margin rounded ; posterior subtruncate ; beaks
large, calyculate, much elevated ; striæ delicate ; green-
ish-yellow ; hinge margin nearly straight ; cardinal teeth
compressed ; laterals slightly elongated. H. 11, W.
15–16, mill.

Station, Schuylkill River, Chester County ; abundant.

S. simile, SAY.

Cyclas similis, Say, Nich. Encyc., Ed. I., 1816.

S. similis. — [Prime.]
Fig. 197.

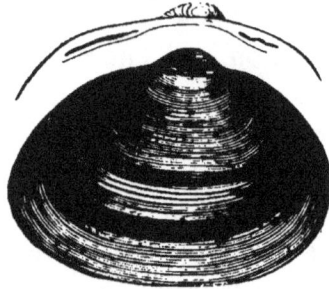

Shell, transverse-oval, more pointed posteriorly, base slightly curved; beaks full, elevated; sulcate, dark chestnut; inside bluish; cardinal teeth small, indistinct; lateral teeth on a line with the primaries, large, prominent. H. 11, W. 17, T. 8, mill.

Station, running streams. Pickering Creek, Chester County; common.

S. striatinum, LAM.

Cyclas striatinum, Lam., An. S. Vert., V., 1818.

S. striatinum. — [Prime.]
Fig. 198.

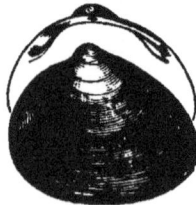

Shell, thin, inequilateral, rounded anteriorly, posteriorly elongated; sulcations slight, or irregular; color light or darker shades of green; interior blue; hinge margin curved; cardinal teeth very small, lateral larger, not prominent. H. 8, W. 11, T. 6, mill.

Station, Brandywine Creek, Schuylkill River, Chester County; abundant.

S. stamineum, CONRAD.
Cyclas stamineum, Conrad, Amer. Jour., XXV., 1834.

S. stamineum. — [Prime.]
Fig. 199.

Shell, stout, oval, full, anterior side abrupt, posterior more distended; beaks very full, and prominent, distant; dark brownish-yellow; interior bluish; strongly striate; hinge margin curved; cardinal teeth nearly obsolete; laterals strong.   H. 9, W. 14, T. 7, mill.
Station, tributaries of the Delaware River.

S. fabalis, PRIME.
Cyclas fabalis, Prime, Bost. Proc., IV., 1851.

S. fabalis. — [Prime.]
Fig. 200.

Shell, transversely oval, almost equilateral, rounded, posterior side rather abrupt; beaks depressed, regularly and moderately sulcated; epidermis, greenish, or straw color; valves thin, blue within; cardinal teeth small; laterals slight, the anterior more elevated, both nearly on a line with the cardinals.   H. 11, W. 14, T. 6, mill.
Station, Susquehanna River and tributaries. Lancaster County; not common.

S. rosaceum, PRIME.

Cyclas rosaceum, Prime, Bost. Proc., IV., 1851.

S. rosaceum. — [Prime.]
Fig. 201.

Shell, small, round-oval, fragile, transparent, sub-equilateral; margins generally rounded; beaks calyculate, approximate, inclined forward; valves very slight, a little convex, fine striate; epidermis reddish-brown, shining; teeth very slight, laterals elongated. H. 4½, W. 6, T. 3½, mill.

Station, Pickering Creek, and Schuylkill River, Chester County ; rare.

S. partumeium, SAY.

Cyclas partumeium, Say, Acad. Nat. Science, Jour. II., 1822.

S. partumeium. — [Prime.]
Fig. 202.

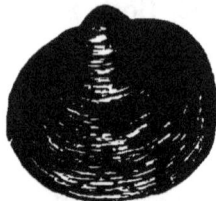

Shell, round-oval, thin, pellucid, nearly equilateral, anterior margin slightly distended; posterior rather abrupt; beaks central, calyculate, approximate; striæ scarcely visible; glossy, greenish horn color; valves moderately convex; hinge margin nearly straight; cardinal teeth strong ; laterals very much elongated. H. 13, W. 11, T. 8, mill.

Station, stagnant ponds on White Clay and Pickering Creeks, Chester County ; abundant.

S. securis, PRIME.

Cyclas securis, Prime, Bost. Proc., IV., 1851.

S. securis. — [Prime.]
Fig. 203.

Shell, thin, rhombic-orbicular, ventricose, subequi-
lateral; margins rounded, basal rather abrupt; beaks
large, calyculate, approximate, inclined forwards; valves
very convex; striæ regular, delicate; greenish or yellow,
glossy; cardinal teeth very small, laterals slight, narrow, elongated. H. 7½, W. 9, T. 6, mill.

Station, ditches and running streams, tributaries of
Pickering Creek. Chester County; abundant.

Genus PISIDIUM, C. PFEIFFER, 1821.

Shell, small, round-oval, inequilateral, anterior side
longer; beaks terminal; cardinal teeth double, or united;
laterals elongated, double in the right valve, single in
the left; ligament on the shorter side.

Animal, lobes of the mantle united posteriorly into a
single short siphon; oral tentacles triangular, elongated;
foot small, tongue-shaped, very extensible.

P. compressum, PRIME, Bost. Proc., IV., 1851.

Pisidium compressum. — [Prime.]
Fig. 204.

9 *

Shell, solid, trigonal, very oblique; anterior side narrower, produced; posterior broad, subtruncate; beaks small, elevated, distant, with a wing-like appendage at summit; regular-striate; yellow, gray, chestnut, mixed; valves, light blue inside; hinge thick; cardinal teeth small, strong, compressed; lateral teeth short, strong, oblique. H. 3½, W. 4, T. 2, mill.

Station, in company with P. variabile. Pickering Creek, Chester County; not common.

P. virginicum, Gmelin.
Tellina virginica, Gmelin, Syst. Nat., 1788.

P. virginicum. — [Prime.]
Fig. 205.

Shell, thick, solid, oblique, anterior side longer, narrower, rounded; posterior broader, subtruncate; base rounded; beaks posterior, large, prominent; interior light blue; epidermis greenish-brown; zoned; hinge margin greatly curved; cardinal teeth strong; laterals strong, and short. H. 7, W. 9, T. 5, mill.

Station, White Clay and Pickering Creeks, Chester County; abundant.

P. variabile, Prime, Bost. Proc., IV., 1851.

P. variabile. — [Prime.]
Fig. 206.

Shell, heavy, oblique, inflated, anterior side longest, narrower, sub-angular; beaks posterior, full, prominent, distant; valves solid; striæ regular, distinct; epidermis glossy, variable, straw, greenish, or yellowish, zoned; hinge margin elevated; cardinals small, united; laterals strong, distinct, short. H. 4½, W. 5, T. 4, mill.

Station, streams and rivulets. Chester County; abundant.

P. abditum, HALDEMAN, Proc. Acad. Nat. Sci., I., 1841.

P. abditum. — [Prime.]
Fig. 207.

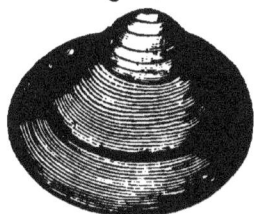

Shell, round-oval, moderately convex, margins well rounded; beaks posterior, small; smooth, straw color; hinge margin nearly straight; cardinal teeth small, separate, the anterior larger, and more prominent; laterals small, not elongated. H. 3½, W. 4, T. 2, mill.

Station, streams, fountains, and rivulets. Chester County; frequent.

GLOSSARY.

Acephala. Animals without heads. Ex., the oyster, mussel, etc.

Adductor (Muscle). The fleshy fibres which close and hold bivalve shells together. They are mostly one or two (Oyster, one ; Unio, two).

Anodonta. Toothless. Ex., the Anodont mussels.

Aperture, or entrance, or mouth, of univalve shells. The mouth of the shell is described as being entire, circular, lunate, semilunate, reniform, angular, sub-quadrate, long, wide, linear, dentate, emarginate, etc. The plane of the aperture. may be longitudinal, transverse, or oblique.

Apex. The tip of the spire, or umbo. It may be acute, obtuse, or sub-spiral.

Base, of a univalve. It rests on the table in the position referred to, and opposed to the apex.

Bivalve. Having two valves. Ex., the clam. They may be compressed, cordiform, cylindrical, inflated, rostrate, or truncate.

Branchial. Relating to the gills.

Buccal plate. The jaw of snails and similar mollusks.

Cardinal tooth. The large, central hinge tooth of Unios and many other shells.

Carinate. Keeled. A longitudinal elevated line.

Carnivorous. Flesh-eating.

Cartilage, of the hinge. An elastic body inside of the ligament of bivalve shells, which is compressed, when the shell
is closed by the contraction of the adductor muscle, and
reopens the shell when the muscle is relaxed.

Cephalic. Of, or relating to the head.

Cicatrix. A scar. The impressions on the inside of bivalves,
caused by the insertion of the adductor muscles. These
are single in the oyster, two in the mussel.

Columella. The central axis round which spiral shells are
coiled.

Conchology. The study of shells, and their inhabitants.

Cordiform. Heart-shaped.

Crenulated. Notched, with rounded teeth.

Dentate. Toothed.

Depressed. With a low, flattened, spire.

Dextral. Shells whose aperture is on the right of the columella.

Discoid. Flat, the whorls of the spire being on the same
level. Ex., Planorbis, etc.

Emarginate. Notched ; appearing as if cut off.

Epidermis. The cuticle, or membrane, which envelops some
species of shells. It varies greatly in texture and appearance.

Equilateral. Having the right and left sides of the valves
equal.

Equivalve. With the two valves equal.

Gasteropod. Belly-footed. A mollusk with a ventral disk, or
foot for walking. Ex., the slugs.

Genus. Kindred. Applied to several things with a common
character. Ex., the snails.

Globose. Spherical.

Helix. A coil, or spiral. Hence the genus Helix.

Hermaphrodite. Applied to those mollusks which recipro-
cally perform the functions of both sexes.

Hiatus. A gape, or opening found in some bivalves when
the shell is closed.

Hinge. The thickened, and mostly toothed, margins of the
valves of a bivalve shell, where they are joined together.
The hinge-teeth are variously named — those under the

umbones are cardinal, or primary; those on either side, lateral, or secondary teeth.

Incurved. Bent, or turned inward.

Inequilateral. Having the sides unequal.

Inequivalve. Having unequal valves.

Involute. Rolled inward.

Jaw. See Buccal plate.

Labia. Lips. The border of the aperture of univalves. The inner lip (columella) may be curved, horizontal, oblique, plaited, or straight. The outer lip may be acute, alated, digitated.

Ligament. The transverse fibres which tie the two valves together. They are generally just behind the umbones.

Limb. The margin of bivalve shells.

Lingual ribbon. See Tongue. Also page 17.

Longitudinal. Lengthwise. In univalves, from the apex; in bivalves, from the umbo to the base.

Mantle. An external thick skin which envelops the bodies of mollusks.

Margined. With a thickened, or colored border.

Microscopic. Very small. Not easily seen without a glass.

Molluscous. Soft. Hence Mollusks, soft animals.

Muscular impressions. The scars on the inside of bivalves, caused by the insertion of the adductor muscles. They are single in the oyster; two, in the unio, etc.

Nacre. The pearly substance of which some shells are formed. It lines the shell of the Unionidæ.

Nodose. Knotty.

Ochreous. Color of yellow ochre.

Olivaceous. A greenish olive color.

Opaque. Impervious to light.

Operculum. A horny scale, or bony plate, attached to the foot of many mollusks, which, when the body is drawn in, closes the shell, like a door.

Ovate. Egg-shaped.

Ovoid. Egg-shaped.

Papillate. Covered with dots, or pimples,

Patulous. Open, gaping.

Pectinated. Toothed, like a comb.

Peristome. The margin of the aperture; the mouth of the shell.

Pillar. The central column which supports the spire.

Plicate. Folded, or plaited.

Pulmonary. Relating to the lungs; animals which breathe.

Reniform. Kidney-shaped.

Reticulated. Forming a network.

Retractile. That can be drawn back. Ex., tentacles of snails.

Reversed. Applied to shells whose spire winds in the contrary way to that which is common. Ex., Physa.

Revolute. Rolled backward.

Rostrate. The elongated canal or beak of univalve shells.

Rostrum. The snout of the animal of univalves.

Rufous. Reddish-brown.

Rugose. Wrinkled.

Scabrous. Rough, harsh, to the feel.

Sinistral. Shells with the aperture on the left side.

Spire. This includes all the volutions but the last, which is known as the body, or body-whorl. The form of the spire may be acute, concave, depressed, discoidal, obtuse, etc.

Striated. Marked with fine lines; longitudinal or transverse.

Sulcated. Deeply furrowed.

Summit. The apex, or tip of the shell.

Suture. The joining of the whorls in the spire. It is more or less deeply impressed, and varies in its character.

Teeth. The aperture of the shell is often furnished with projections from its inner margin, which are called teeth. Ex., Triodopsis.

Teeth, of molluscous animals. See Tongue.

Tentacles. The feelers of snails, etc.

Tongue, of the mollusca, is a horny, strap-shaped organ placed within the mouth, and armed with very numerous microscopic teeth, arranged accurately in rows both transverse and longitudinal, to the number, in some species, of many thousands. It serves to rasp off their food.

Transverse. Crosswise. The measure, or width of a shell horizontally from right to left.

Truncate. Appearing as if abruptly cut off.

Tubercle. A small wart or pimple.

Turbinate. Top-shaped.

Turgid. Full, as if swollen.

Turreted. Extended to a long point.

Umbilicus. The perforation in the pillar, at the base of the shell. It may be deep, narrow, wide, etc.

Umbo. Umbones. The protuberance seen on the valves of most bivalve shells, near the hinge. The part first formed. It presents various interesting characters.

Undulated. Waved.

Univalve. Composed of one valve, or piece. They are mostly spiral, sometimes conical, as in Ancylus.

Valve. Signifies one of the valves of a bivalve shell. The two valves constitute only one shell.

Varices. Longitudinal ribs, crossing the whorls of univalves.

Venter. The belly. The right hand portion of the body whorl of univalves ; and the base, or lower edge of bivalves.

Ventricose. Inflated, bulged in the middle.

Verrucose. Warty.

Viscid. Glutinous. Ex., the slugs.

Whorl. One turn of a spiral shell.

10

INDEX OF GENERA, AND SPECIES.

THE END.